FINDING OIL

FINDING OIL

The Nature of Petroleum Geology, 1859–1920

Brian Frehner

UNIVERSITY OF NEBRASKA PRESS

LINCOLN AND LONDON

Funding for the index has been
provided by the William P. Clements
Center for Southwest Studies.

Library of Congress
Cataloging-in-Publication Data
Frehner, Brian.
Finding oil: the nature of petroleum
geology, 1859–1920 / Brian Frehner.
p. cm.   Includes bibliographical
references and index.
ISBN 978-0-8032-3486-4 (cloth: alk. paper)
1. Petroleum industry and trade — United
States — History — 19th century.
2. Petroleum industry and trade — United
States — History — 20th century. I. Title.
HD9565.F727 2011
333.8'232097309034 — dc22
2011008716

Set in Arno by Bob Reitz.
Designed by Nathan Putens.

To Celeste

Contents

Illustrations

Acknowledgments

Recounting all the people who have helped me navigate my journey to becoming a historian and to producing this manuscript is humbling. The first person I encountered in the historical profession is also the first person I have to thank, the late Hal Rothman. Hal had just arrived at the University of Nevada–Las Vegas when I decided to leave Los Angeles and return to my hometown to begin graduate school. He guided me as a mentor and provided numerous employment opportunities as an adjunct instructor and as a public historian while I was at UNLV and after finishing my MA. Studying under Hal and working for him taught me more than I appreciated at the time. As I would later learn, Hal knew virtually everyone in the historical profession, and his larger-than-life personality left a strong and lasting impression. We lost a memorable figure with his untimely passing. He was always in my corner, and I miss him. I would also like to thank other professors who provided their support and assistance while at UNLV, including Eugene Moehring, Sue Fawn Chung, and Andy Fry.

I also have to thank Hal for directing me to Arizona State University, where I met Albert Hurtado, who began directing my PhD work and who I was lucky enough to follow to the University of Oklahoma, where I completed my degree. I have learned much from working with Al and continue to benefit from his sage advice and model of professionalism. My debt to him is greater than I can ever repay, and my friendship with him and his wife, Jean, is one of the unexpected joys that studying history has presented me. I feel lucky to have encountered them and look forward to their continued friendship. While at ASU, a fellow student, Jaime Aguila, presented memorable distractions from class work, and he and his wife, Holly, have become lasting friends. I also must thank Donald J. Pisani, who cochaired my dissertation committee with Al at the University of Oklahoma and read many drafts of the dissertation and manuscript. Don continually challenged me to "find the story," and he can judge whether I succeeded or failed. Other professors at the University of Oklahoma who helped me and who I now thank include Robert Griswold, David Levy, Roberta Magnusson, Paul Gilje, Rich Hamerla, Joshua Piker, Ray Canoy, and Catherine Kelly.

I had the great fortune to spend a year as a postdoctoral fellow at the Clements Center for Southwest Studies, and that experience presented me with opportunities to meet and interact with an amazing cast of talented and supportive people. Heading the list, of course, is the late David Weber, who I thank for his guidance, assistance, and the kind spirit he displayed in all of my encounters with him. I think that all who knew him recognize that David was not only a great scholar but an exceptional person who possessed a charismatic presence and an instantly likable glow in his eyes. I feel privileged to have spent a year at the institution he built. While at the Clements Center, I learned that much of its dynamic and energetic spirit was due to the presence of Sherry Smith. She has been generous with her time, and I have learned much from her about balancing history with life. The Clements Center is so successful because David knew how to surround himself with smart and vibrant people like Sherry, and he found another in Benjamin Johnson. I am grateful to Ben and his wife, Michelle Nickerson, for reading my manuscript, for their professional guidance, but mostly for their friendship. Andrea Boardman and Ruth Ann Elmore gave their time and consideration generously to the endless details that make the fellows' lives easier and for that I thank them. I found at the Clements Center a great friend in Andrew Graybill, who along with his wonderful family provided evening breaks from work with food, laughter, and great company. The manuscript workshop is the highlight of the fellow's experience at the Clements Center, and I thank all of the following who attended: John McNeill, Phillip Scranton, Laura Hernández-Ehrisman, David Blackwell, Suzanne Bost, Dennis Cordell, Alicia Dewey, John Mears, Houston Mount, Robert Righter, and Rajani Sudan.

I was invited to participate in various symposia at which I presented portions of chapters, received valuable feedback, and met many kind and generous people. I am very grateful to Karen Merrill for reading different incarnations of the manuscript and for all of her guidance as well as her invitation to participate in the "World Oil Frontiers" symposium at the Howard R. Lamar Center for the Study of Frontiers and Borders at Yale University. Bill Deverell invited me to participate in "Under the West," a symposium at the Huntington Library, where he was a most gracious host. I greatly appreciate Jeremy Vetter's invitation to present a paper on the topic of lay science at the Max Planck Institute in Berlin, where I encountered an international group of scholars with similar interests.

Many archivists and librarians throughout the United States assisted me with the research for this manuscript. I would like to thank John Lovett at the Western History Collections, University of Oklahoma; the DeGolyer Library, Southern Methodist University; the American Philosophical Society; the Hagley Museum and Library; the American Heritage Center, University of Wyoming; and the Center for American History, University of Texas–Austin.

I was fortunate to receive research funding from the history departments at Oklahoma State University and the University of Oklahoma. My colleagues at Oklahoma State have been very supportive, as have two department chairs, Elizabeth Williams and Mike Logan. I appreciate the assistance provided by the following research fellowships: the Bernard L. Majewski Fellowship at the American Heritage Center, University of Wyoming; Clements Center–DeGolyer Library Research

Travel Grant, Southern Methodist University; the Hagley Center for the History of Business, Technology, and Society travel grant; and the Bea Mantooth Estep History of Oklahoma Dissertation Scholarship, University of Oklahoma.

I am grateful for all of the support my family has provided over the years, and I thank them all and hope that I have made them proud. I am also thankful for the serendipitous choice of buying a home next to my favorite oboist, Celeste, and am grateful every day for her patience, support, and presence. I appreciate all that she does and love her for reasons she knows best.

FINDING OIL

Introduction

Shortly after walking over the dry west Texas plains, Jett Rink knelt on the ground while squeezing handfuls of oil-soaked dirt through his fingers and gazed in amazement at the black crude slowly bubbling to the surface. Later, Rink stood atop a cable tool drilling rig when a loud noise caught his attention. The black crude that had merely bubbled to the surface began to emit an awesome roar as it erupted from the hole Rink punctured in the earth. He stepped back to behold the spectacle he had created, as oil spewed from the earth and rained down on him. He held both hands in the air as if to thank Mother Earth for her beneficence, and jumped up and down to celebrate his good fortune.

The image of a gusher is a powerful symbol in the history of the American Southwest. Dramatized by James Dean in the movie version of Edna Ferber's novel *Giant,* this scene played out repeatedly throughout the early twentieth century in the history of oil-rich states such as Kansas, Oklahoma, Texas, and California. Captivating and dramatic, a gusher represented a visual image of nature's bounty spewing forth uncontrolled and seemingly uncontrollable. The image was powerful but far from simplistic: it held different meanings for the prospectors who found gushers as the oil industry grew and matured. For example, an oil prospector like Jett Rink might see a gusher as a symbol of great wealth, a fortune in the making, while to another prospector the gusher symbolized profligate waste and technological incompetence. Oil prospectors expressed these different views about oil over time. However they conceived of gushers, all prospectors strove to translate the geological forces governing oil into other forms of power — economic, intellectual, and cultural — within this emerging industry. This book recognizes the diversity of their views but shows that important similarities existed in the kinds of knowledge cultivated by the most successful prospectors.

The central argument of this book is that oil prospectors struggled for cultural, intellectual, and professional authority — over both nature and their peers — from 1859 to 1920. Throughout the oil industry's early history, multiple people from varied class, educational, and professional backgrounds vied for the authority to determine where oil resided. Despite the towering presence of a figure like John D. Rockefeller as the quintessential "oil man," prospectors made up a diverse lot who saw themselves, their interests, and their relationships with nature in different ways primarily through their work. Oil men established relationships with nature through their work in order to harness geological forces and unleash oil in a rushing flow, but they encountered and understood geological forces differently. At the center of the relationships they formed with nature lay a struggle for power, position, and prestige within the oil industry and, for some, within the scientific community. Certainly economic gain motivated many prospectors, but the knowledge they cultivated and articulated about oil and its character bestowed upon them intellectual and cultural power too. Finding (and failing to find) oil through physical and intellectual work taught prospectors knowledge that they built upon as the industry evolved, but this process often unfolded irrationally and with intense conflicts among people who disagreed about how and where to find oil.

The idea that people within the oil industry engaged in a quest for power hardly seems novel given oil's central importance throughout the twentieth century, but this book complicates that story by arguing that the prize prospectors sought constituted power that did not always equate to financial gain. Clearly discoverers of large oil deposits potentially stood to amass fortunes, and this prospect alone motivated many to explore for subterranean riches. Explaining the behavior of all prospectors on the basis of financial gain alone, however, illuminates their motivations no more clearly than does the argument that fish swim because they live in the water. Indeed, one very popular book on the history of the oil industry argues that "oil has meant mastery" and it was this "quest for mastery" throughout the twentieth century that constituted an "epic quest" for "oil, money, and power."[1] Indeed, the men who formulated the knowledge for locating oil (and they were mostly men) successfully applied their learning to uncover vast stores of oil, and they profoundly and radically changed the world through their discoveries in the United States. Their story is significant because the power and mastery they wielded took many forms that prevented prospectors from achieving a consensus in their supposed mastery over the geology that housed oil. This book explains that some prospectors saw power as the work that they expended exploring and drilling into landscapes. Yet others saw power more in terms of an intellectual activity from which they derived geological theories from field work, using those theories to explain how and where oil accumulated. In the end finding oil required both physical labor and intellectual theorizing; prospectors possessed different cultural orientations toward their work and toward nature that they could not always reconcile.

With this story I hope to join an emerging conversation among scholars who see environmental history as a field in which nature and culture coexist in a tangled dialectic rather than as discrete categories of analysis. More environmental historians recently have taken a "cultural turn" by telling stories of hybrid landscapes that blur the lines between a supposedly pure and pristine nature and manifestations of human culture.[2] Americans related to, thought about, and interacted with nature as laborers, scientists, and industrialists, to name only a few occupations that gave rise to complex and often competing discourses about the worlds they inhabited.[3] Hybrid landscapes show that the intersections between culture and nature changed depending upon the times and places in which they occurred. Some of the literature on this topic dispenses with easy categorizations of social or economic behavior by showing that production and consumption of nature reflected historically specific contexts and often constituted a paradox. The very terms "production" and "consumption" take on different meanings depending on when and where they occurred, and these differences thus complicate the social and economic activities that shaped cultural experiences in nature. For example, miners who dug into mountainsides produced gold but simultaneously consumed nature by picking wild berries or eating canned and packaged food to feed themselves.[4] Tourists who culturally consumed nature inside national parks simultaneously produced revenue for surrounding communities through their wilderness encounters. This monograph will present numerous oil producers whose highly personal (if not intimate) encounters with landscapes reveal cultural conceptions of nature that grew from relationships they fashioned through their intellects and physical labor. What is at stake for the people in all of these histories is the power to define what constitutes nature.[5]

At the center of my analysis is the idea that the geological processes that created and trapped oil beneath the ground constituted particular environmental contexts that challenged prospectors to encounter nature both physically and intellectually in order to locate, extract, and control the resource they sought. Apart from a gusher's symbolic power, an uncontrolled oil well illustrated in dramatic, visual form the fugitive qualities of this natural resource, physical attributes that influenced how people searched for, confronted, and attempted to control oil. Not all wells gushed, but even those that seeped or flowed revealed oil geology and suggested technology for locating and extracting the elusive viscous resource. Other natural resources such as coal, timber, water, and native grasses each possessed unique qualities that shaped how people attempted to appropriate and eventually conserve

1. Iconic image of a gushing oil well. Celebrated and cherished by some, quickly suppressed by others. This is the Lakeview No. 2 gusher in Kern County, California, May 20, 1914. Courtesy of United States Geological Survey.

them. Unlike forests, rivers, and grasslands, however, the great bulk of oil reserves lay hidden from human view by complex and varied geological formations. This book is about how people attempted to access that subterranean world and the ideas they conceived of where oil lay, ideas that sometimes led them to oil but often led them into disagreements and conflicts with one another. The physical qualities of oil and the varied geological conditions in which it resided thus shaped human efforts to appropriate the resource. Just as continental sheets of ice and raging forest fires influenced humans' responses to them, oil's physical properties conditioned humans' responses to it.[6]

Prospectors debated oil and its geology because the work they performed mediated their relationships with nature in highly individualistic ways. Battles waged over oil-finding technique reflected a broader debate over prospectors' relationships to nature through their labor. Work is one of the primary means for human beings to know nature.[7] Few humans throughout time and across the globe have escaped the need to perform work for food, clothing, or shelter. Whether hunting for food, tilling the soil, or prospecting for oil, people who labor simultaneously alter the natural world and learn about nature through the physical energy they exert. What they know they learn through their bodies, which serve as conduits that absorb nature's dictates to shiver, sweat, gaze, peer, or listen in order to track a moose, plough a field, or drill a hole into the earth. We understand our relationship to nature when we understand its relationship to the work our bodies perform. People also make sense of their surroundings by filtering their experiences through cultural lenses such as language and epistemological orientations that process and order their relationships to nature, connections constantly in flux as new sensory inputs require their refashioning.[8] The work people perform can alter environments, but this labor also functions as the medium through which the environment can change people. Like all relationships, people and environment interact continually, and each participates in an ongoing dialogue in which one affects the other.

Oil prospectors used their bodies and minds to engage with nature while conducting field work. Some prospectors found oil by relying on their eyes, ears, feet, and hands while traversing landscapes. Others saw field work as an opportunity to formulate philosophical speculations about geological principles useful for finding oil and answering larger questions about a landscape's form, age, and relationship to surrounding topography. The experience of gathering local knowledge in these ways could lead to oil, but it also imparted an environmental ethic or sense of place that varied depending upon when and where prospecting took place. Because individual prospectors employed varying degrees and combinations of physical and intellectual work, their experiences in nature ranged the gamut from an appreciation for nature's aesthetic beauty to a supposedly detached objectivity that facilitated geological theorizing.

The different kinds of knowledge prospectors generated from their work gave rise to intense debates over what constituted reliable and trustworthy geological information. I intend throughout this study to complicate simple polarities between "objective" and "subjective" knowledge in order to highlight how the "science" of petroleum geology resulted from an ongoing collective process in which many individuals participated.[9] Relationships that prospectors formed with nature could at times be highly individualistic or even eccentric, presenting opportunities for some to contest, disprove, or revise another's theories for locating oil. Understanding that knowledge coalesced through these negotiations illuminates the complicated matrix between nature and different constituencies, but there is also a moral dimension to the relationships prospectors fashioned with nature and the knowledge they produced.[10] People generate knowledge because of the need to carry out a practical activity. When knowledge achieves this goal, it embodies a collective good and possesses a moral component because nonexperts must rely upon others and trust in others' observations of nature in order to make decisions that would affect their lives. Thus knowledge has natural *and* social components.[11] Many people practiced the practical task of searching for oil, and in order to find oil, they had to generate knowledge that warranted a strong consensus to attract investors' interest or to justify the time and expenditure of drilling a hole. The knowledge generated for determining where to drill that hole grew out of people's willingness to trust the knowledge prospectors generated for locating the best possible site.

This book explores the process of how people identified trustworthy agents but more importantly how those agents developed and demonstrated their trustworthiness as professional geologists and accrued authority over a body of knowledge that expanded over the timeframe of this study—and continues to grow in the present. As consumers of oil living in the world today, each of us must trust scientific experts working for oil companies who proclaim the safety of their production methods or that the costs involved in finding oil justify prices we pay at the pump. Geologists still play a central role in finding oil, but this book argues that they fought long and hard to win the public's trust and the confidence of investors and executives within the oil industry.

Casting knowledge as a highly contested by-product of different practitioners' work in nature during the timeframe 1860 to 1920 illuminates how professional scientists and engineers only gradually imposed their practices on the oil industry.[12] To gain acceptance, geologists and engineers borrowed and built upon knowledge generated by lay practitioners whose physical and experiential encounters revealed valuable information about geology in local contexts. Some petroleum geologists recognized they could enhance their authority by extrapolating theoretical abstractions from local knowledge and applying these local practices universally in order to commodify oil in other locales. Lay practitioners, on the other hand, facilitated the cultural and bureaucratic enshrinement of objective and scientific knowledge in the oil industry as much as professional scientists but most often at the local level. Lay practitioners did not disappear at the time this study ends, but their role in finding oil had grown significantly more marginal as petroleum geologists staked their claim to universal knowledge and expertise.

Prospectors never fitted easily or neatly into categories because their methods often overlapped, but two of the most prominent types to emerge in the oil industry were the practical oil man and the professional geologist. Jett Rink provides one example of a practical man because he prospected alone and had not received formal geological training. Conversely, geologists typically educated themselves at universities, where they formally studied geological principles, but they also supplemented this training with experience working in the field or for an oil company.

Although I use the above labels throughout this study, I realize the appellations potentially prove troublesome because they overstate the differences between the two groups and obscure similarities. Prospectors looked for oil with a range of approaches, but those who found oil most often employed very similar forms of knowledge. Thus the labels function as ideal types, with each type potentially generating knowledge that succeeded or failed to yield oil. Practical oil men functioned as craftspeople and technicians who gathered knowledge to perform the material task of finding oil and selling it to the highest bidder. The knowledge they generated possessed a vernacular character and reflected ways of knowing the natural world mostly situated in the physical labor they performed. Geologists also wanted to find oil and cultivated local knowledge, but they aspired simultaneously to formulate innovative theoretical insights that built upon existing geological abstractions in order to accrue status and prestige as members of a professional scientific community. Perceived differences between these two types often accelerated and exacerbated contests for power and authority among them. They often dismissively caricatured and oversimplified each other and their methodologies as either too narrowly based on local environments or too theoretical, abstract, incomprehensible, and therefore impractical. The tension between the diverse array of prospectors and their methods propelled and sometimes inhibited formulation of the fundamental tenets of petroleum geology and engineering, which the oil industry did not universally adopt until the second decade of the twentieth century.

If the methods of different practitioners overlapped, how can we understand differences between them and the categories they used to describe themselves and their work? First, we must recognize that categories matter greatly to the people who create them. Categories people use to classify themselves, others, and the work performed by each originate from decisions to segment the world both spatially and temporally, to create a set of literal or metaphorical boxes in which they place their work and production of knowledge.[13] We should take seriously the labels people create to identify themselves and their work, but we must also remember that each category either valorizes or silences a particular point of view.[14] When people make choices, they exercise power. The choice to valorize or silence another historical actor means that categories reflect moral and ethical agendas.[15] When a prospector called himself a practical man, he enshrined his knowledge, castigated geologists, and defined oil prospecting as work befitting only the male gender. Similarly, practitioners who labeled themselves geologists valorized their work as translocal knowledge that encompassed study of broadly defined earth processes and differentiated themselves from prospectors, whom they considered merely provincial rather than practical.

Regardless of how they labeled themselves or each other, a common desire to find oil often brought prospectors together even if their relationships did not always prove easy. People disagreed frequently about how to find oil, but they learned through a collective process that unfolded gradually and incrementally and through trial-and-error attempts. Human beings learn though experience even if they do not always succeed at their endeavors, but few learn in isolation or without guidance from a parent, mentor, coworker, or friend. Observing an expert at work may also teach a young apprentice and impart a particular set of skills. What these methods have in common is the idea that people are social creatures who learn in collectivities or in collaboration with one another. Learning collectively requires people to form relationships that allow them to share a repertoire of resources such as stories, tools, or techniques for accomplishing particular tasks.[16] An ideal learning community consists of people who share a similar concern, passion, or practice for achieving a common goal.[17] What distinguishes communities of individuals who teach one another from a mere interest group or social club is the need for members to function as practitioners, to perform an activity that requires time, attention, and sustained interaction with others who possess varying levels of expertise. Examples of such communities might include a tribe learning to subsist on a new resource base, musicians crafting new music, engineers fixing a bridge, or prospectors searching for oil. Finding oil was, in part, a social process that required interpersonal communication and relationship building.

When people fail to communicate and cannot understand one another, their relationships suffer, and even harmonious communities fracture. To understand how people learned where to drill for oil, we must not romanticize prospectors as a community of practitioners whose members coexisted in solidarity, harmony, and consensus. Indeed, fractures occurred repeatedly among oil prospectors, and these breaks will reappear throughout this narrative because they reveal struggles for power and authority among people who had different understandings of how their work mediated their relationships with nature. Furthermore, these disagreements illuminate the contested nature of the knowledge that made up petroleum geology. The science and technology oil prospectors fashioned to find and produce oil grew out of contested relationships, decades in the making, in which prospectors agreed and disagreed about their practices. These disagreements sometimes facilitated understanding of geological theories, but they also inhibited the formulation of these theories and prevented their acceptance by the industry. Innovation and efficiency in exploration occurred within the oil industry not necessarily by unanimity and consensus but from hotly fought turf battles in which practitioners defended their intellectual and cultural terrain. Chapter 1 depicts some of these battles over the chronological scope of this study and argues that vernacular prospectors frequently established their authority as oil finders and often at the expense of geologists.

Geologists worked in the oil industry throughout the late nineteenth century mostly as consultants and generated useful scientific knowledge, but their influence remained limited until the early twentieth century.[18] Their work on state geological surveys produced a unique blend of culture and nature that did not always translate into the most efficient knowledge for locating oil. Directors of geological surveys and their assistants gained valuable experience conducting field work in order to meet taxpayers' demands for information leading to natural resources. Surveys were highly politicized undertakings when the knowledge scientists generated proved too arcane for layman to understand, but these surveys offered geologists opportunities to accrue authority and power over geological knowledge even if they did not always realize that goal. Chapter 2 explores how opportunities for scientific and professional authority precipitated conflicts during the nineteenth century between geologists who directed Pennsylvania's first and second geological surveys and the assistants they employed. Their disagreements did not always specifically involve oil but concerned issues of professional authority and geological expertise. Tensions flared over the question of who could properly claim credit for knowledge the survey generated — assistants, who performed much of the field work, or its director, who assimilated the findings? Claims for authority led geologists to fight among themselves and denigrate practical men's prospecting theories. The most authoritative knowledge regarding petroleum geology resulted from collaborations initiated by geologist John F. Carll, who gathered factual data from practical men's drilling logs to construct a sound geological theory explaining how and where Pennsylvania's subsurface geology trapped oil. As demonstrated in chapter 3, practical men retained authority well into the second decade of the twentieth century. Tom Slick, a practical oil man who headed west from Pennsylvania, discovered the immensely productive Cushing, Oklahoma, oil field even though he had no formal training as a geologist. Despite Slick's success, the authority he and other practical men accrued began to wane as the twentieth century dawned, and the oil frontier shifted to the southern plains.

Geologists began to displace practical men as oil-finding authorities during the first two decades of the twentieth century by building institutional power within universities and surveys and using these forums for creating and controlling knowledge the oil industry wanted while simultaneously advancing their professional authority and prestige. Much of chapter 4 discusses how Charles N. Gould, although not the best oil prospector, accrued institutional power as a university geology professor in Oklahoma and as director of the state's geological survey. He also accessed power at the federal level by lobbying the United States Geological Survey for assistance in locating Oklahoma's mineral resources. Petroleum geology began to coalesce as a formal discipline presided over by geologists who systematized their field work practices into methods for generating theories of oil accumulation and documenting these theories with surface and subsurface maps. Collaboration still produced the most important geological knowledge, but the collaborators increasingly possessed university degrees, trained under the purview of public institutions, and accrued authority for knowing nature in ways private industry greatly valued.

Private institutions in the form of large integrated oil companies increasingly recognized that geological expertise offered predictability in finding oil and began hiring geologists from universities and surveys for which they practiced the art and science of mapping the varied geological conditions that trapped oil. Probably more than any other single figure, Henry L. Doherty of Cities Service paved the way for the acceptance of geology in the oil industry. Chapter 5 shows that oil companies' interest in geological exploration sprang from Doherty's efforts to fashion a context in which science, technology, and nature interacted as a "technological system." Doherty's bold decision to build a permanent research staff comprised of over two hundred geologists demonstrated that geology had its place within the integrated structure the oil industry had been building. Like transitional figures in other industries, Doherty bridged a gap in the oil business between the era of "heroic independent invention" in the late 1800s and the period of organized industrial research that coalesced in the early twentieth century.[19] Although John D. Rockefeller had assembled an integrated oil company before Doherty entered the industry, Standard Oil focused its efforts on refining, transportation, and marketing and purchased the oil it acquired from large and small prospectors throughout the country. Doherty recognized that prospecting for oil and its extraction from the ground remained unsystematic, decentralized endeavors and moved to organize and rationalize these activities. This move meant hiring university-educated geologists and engineers whose knowledge seemed more tangible, predictable, and therefore reliable than the tacit and intuitive approach practical men employed with great success. He committed more resources than any previous oil company to searching for oil through the application of geological principles. Other companies followed suit, and throughout the 1920s the industry located so much oil that overproduction, which had been a problem since Drake's well in 1859, returned with a vengeance.

The central aim of this book, then, is to provide a historical context for different oil prospectors and to explain that the evolution and acceptance of petroleum geology within the oil industry grew out of relationships prospectors formed with each other and the natural environment. Work complicated the relationship between culture and nature but so too did interprofessional competition among different practitioners. Many scholars have written and theorized extensively on professionalization. They have seen the process as growing through a series of predetermined stages that represent either a positive story of knowledge as triumphant practice or a less sanguine narrative in which professionals behaved as servants of power working in the interests of corporate monopolies. The best histories on professionalization view professionals' self-definitions and their intellectual activity with skepticism and focus on the interstices between those who do and do not qualify as professional. This perspective casts professionalization as a study in the construction of cultural authority and power relations among different practitioners. By considering professionalization as the cultural negotiation of power, heretofore marginalized historical actors find their way into history as subjects who potentially wielded power in their own right.[20]

Professionalization was not a static process in which geological knowledge triumphed or in which geologists functioned solely as tools of monopolistic oil companies. Rather, the group of practitioners who labeled themselves geologists and coalesced as a distinct profession by 1920 disputed jurisdictional boundaries at the local and national levels. They did so by attempting to dominate outsiders, or practical men, whose prospecting methods they considered attacks on their control of knowledge. Petroleum geology's professionalization involved association formation, licensing, and development of ethics codes, but I argue that the real story was one of competition and contestation for control of knowledge and of nature. As one scholar put it, "It is the history of jurisdictional disputes that is the real, the determining history of the professions."[21] Elucidating the competition that took place among different practitioners explains why the organizational form of the oil industry evolved from practical participation in the nineteenth century to geological research departments staffed almost entirely with university-trained geologists by 1920.

Science and technology play prominent roles in this story because they mediated both practical men and geologists' relationships to nature. Historians attempting to understand modern science no longer believe they can trace universal concepts, theories, and practices back through time to their origin in a supposed Scientific Revolution.[22] Over time humans have attempted to know nature in such a variety of ways that concepts familiar to modern science such as experiment, observation, and objectivity acquired a unique historicity specific to time and place.[23] Rather than a proscribed body of knowledge, science represented an accumulation of various practices situated in local environments in which training, socialization, and knowledge formation reflected the context of a given locality.[24] Such was the case in Pennsylvania, where this story begins, and on the southern plains, where the story ends.

Although California will appear at times in this narrative, the bulk of the story involving petroleum geology's relationship to private industry in that state constitutes another story for another time. Particularly before 1920, California's geographical distance from the oil-rich southern plains region meant that the state's oil industry grew in relative isolation, and while producers instituted geology, they accessed markets overseas by loading oil onto ships rather than sending it overland through pipelines.[25] California's marketing of crude gave rise to a distinctive political economy in which state and local politics structured the oil industry in that time and place along with science and environment but making for a much different story than unfolded on the southern plains.[26]

State and local politics mattered to varying degrees in all oil-producing states, but this monograph aims primarily to examine the contingency of science as producers migrated from Pennsylvania to the southern plains environment from 1859 to 1920. Science did not arrive as a monolithic entity in which development occurred inevitably and triumphantly on the basis of a single experimental method.[27] Similarly, technology did not consist merely of material artifacts or utilitarian phenomena geared toward optimal efficiency but was a socially and culturally constructed tool shaped by a range of people, circumstances, and contexts.[28] Neither science nor technology advanced in a simple linear fashion but intermixed and originated within discrete cultural and environmental contexts to mediate people's relationships to nature in historically specific ways.[29]

If prospectors operated within a mélange of culturally constructed nature, science, and technology, as this study argues, how can we identify and categorize the knowledge called petroleum geology that the oil industry embraced when alternatives frequently presented themselves? No single category offers an answer to that question. Rather, we must consider a range of both human and nonhuman actors as part of an ongoing and evolving network in which innovation occurred.[30] In the southern plains environment there arose an actor-network consisting of people and their scientific and technological innovations that operated as a holistic system in which human and nonhuman actants defined one another based upon the nature of their exchanges and interrelationships.[31] In short, nature and culture shaped one another. Sociologists and philosophers of science and technology who originated theories of human/environmental interchange provided a useful conceptual framework that parallels the work of environmental historians who also strive to capture the complexities of how nature and culture interact.

Prospectors who located and extracted oil offer case studies of how people understood nature through their labor, and the cultural creations that we sometimes label science and technology mediated their efforts. Differing geological configurations participated in the dialectics between humans and landscapes, challenged prospectors, and sometimes foiled their best efforts to dominate nature. Environmental historians have long contended that "we cannot understand human history without natural history and we cannot understand natural history without human history."[32] At its core this book is a story about the intersection of human and natural history. A blasting gusher illustrates that intersection perfectly. Volumes of oil spewing hundreds of feet skyward struck some prospectors as nature in its purest form, yet if not for a human element the geology containing those gushers would never have been disturbed.

1 Local Knowledge

1

Vernacular Authority
in the Oil Field

In 1940 the Union Oil Company of California ran an advertisement in the pages of *Fortune* magazine boasting of scientific innovations that eliminated a prominent figure from the oil industry, a man with a little black box, often known as a doodlebug. To townsfolk he had become a familiar face in the oil fields and was recognizable by his intent focus on the "mysterious black box" that promised to detect oil beneath the ground. Although doodlebugs prospered until about 1898, the ad says that they disappeared shortly afterward because Union eradicated them as the first oil company to establish a department dedicated to studying the field of geology and hiring practitioners to find oil. The doodlebug disappeared because "science killed his career."[1]

The advertisement tells a story in which prospectors relying upon folk knowledge used "unscientific" methods to locate oil, and it suggests that doodlebugs deceived people who hoped to find oil. However, the ad only hints at another part of the story, the contested nature of the local knowledge used to locate oil and how different types of prospectors claimed authority for generating that knowledge. Doodlebugs were one of several types of oil finders who fell under the banner of "vernacular prospectors" and relied upon a variety of techniques, some of which involved surveying the landscape and some that did not.[2] Many vernacular prospectors who surveyed the landscape gathered valuable knowledge about how and where to find oil and met with great success long before the discipline of petroleum geology coalesced in approximately the second decade of the twentieth century. They interacted with a diverse array of environments and encountered different geological formations, sometimes consciously and sometimes subconsciously, in order to cultivate local knowledge that informed their decisions about the best sites to drill for oil.

The Union Oil advertisement posits a view of science as a rational process that many would recognize, but it greatly oversimplifies how people found oil, why they searched, and how petroleum geology coalesced. Geologists certainly enhanced their stature and expertise as the oil industry grew, but the trajectory made clear in the advertisement, from pseudoscientific quackery to triumph of organized, rational, and "scientific" methods, represented a moral narrative casting geologists as heroes and vernacular prospectors as villains. If not for scientific innovations introduced by the likes of Union Oil, the ad suggests, the United States may have fallen prey to the skullduggery of cheats and swindlers.[3] A number of different constituencies contributed to the body of knowledge large corporations used for locating oil, and some of those contributors included the very doodlebugs derided in the advertisement. This chapter will survey three categories of people who prospected for oil in the United States from 1859 to 1920 — charismatics, practical men, and geologists — and it will describe their methods and consider how their methods for cultivating local knowledge overlapped.

LOCAL KNOWLEDGE

The history of oil prospecting throughout the nineteenth and early twentieth centuries provides an excellent example of how oil finding constituted a socially constructed process that changed over time and place. Throughout these decades, science contributed greatly to industrial development as large-scale business enterprises created research and development departments and staffed them with university-trained experts in order to generate knowledge that businesses could translate into profit.[4] The very idea of science, however, underwent a great transformation throughout the course of the nineteenth century, and not until the end of this period would the term begin to acquire its modern meaning of an objective enterprise based upon original research.[5] Scientists who specialized in geology played an important role in the oil industry, but laymen also fashioned oil-finding methods and often had as much or more success than university-trained scientists.[6] In the oil industry craft knowledge, tinkering, and learning through hands-on experience greatly facilitated industrial growth and contributed to the definition of what constituted geological science.[7]

At the core of the methods and techniques laymen fashioned to find oil lay a body of knowledge tacit in character and cultivated in local environments. Throughout human prehistory the transfer of technology between different people took place on the basis of nonexplicit learning.[8] The term "tacit knowledge" refers to ways of knowing that are intuitive and instinctive and that do not lend themselves to articulation or codification.[9] Much of the knowledge that laymen cultivated to prospect for oil was tacit but also local in character.[10] The term "local knowledge" will refer to the sets of beliefs and judgments prospectors made conducting field work in order to study the geology of a particular locale and that they used to inform their ideas of where to drill. This chapter focuses upon two groups of vernacular prospectors, referred to as "the geological underworld" by one historian.[11] The first group of vernacular oil finders I call charismatics, to borrow a phrase from Max Weber; they claimed to possess psychic and supernatural abilities to locate oil. The second group, practical oil men, disavowed supernatural powers, luck, or good fortune in favor of a more rational and disciplined method that involved surveying the landscape for physical signs of oil.[12] Members of both groups cultivated local knowledge, although some performed this task better than others. Local knowledge proved central to Euro-Americans' practices of exploring, inhabiting, and taking possession of areas they settled.[13] This process repeated itself on the oil frontier, where vernacular prospectors used local knowledge to begin to comprehend the lands under which much of the nation's oil resided.

The oil industry was born in Pennsylvania in 1859 when Edwin T. Drake initiated the first commercial oil well. People throughout the world had used oil for centuries for various purposes, but at the time of Drake's well, Americans did not use oil primarily as an energy source. No systematic method for finding oil existed throughout most of the nineteenth century, and prospectors relied upon a spectrum of techniques to find it. Vernacular prospectors frequently and consistently found oil throughout history by cultivating knowledge with their bodies. For those who surveyed the landscape, searching for oil was primarily a sensory experience. Locating oil required prospectors to encounter the landscape physically by *looking* for it with their eyes, *smelling* it with their noses, and *touching* it with their hands. Finding oil in this manner remained possible in areas where oil had not been extracted and still bubbled out of the ground to create seeps that the eyes, nose, and hands could identify. A seep exuded anywhere from a few drops to a few barrels of oil.[14] Oil seeped into creeks and streams, intermixing with the water and impressing the eyes as it rotated on the surface in rainbow swirls. Because the oil seeps and associated natural gas resided close to the surface, they emitted distinctive sounds and smells signaling that gas and oil might be lurking nearby, perhaps even directly beneath the feet.[15] Gas also revealed itself to prospectors' eyes when it bubbled to the surface of a stream or to their ears when it issued forth in a volume large enough to emit a whistling noise.[16] Surface indications other than oil and gas seeps included asphalt veins and oil-impregnated outcrops, all phenomena that led prospectors to the first commercial oil fields in America and guided the search for oil throughout the industry's first fifty years.[17]

When undiscovered oil did not seep to the surface, prospectors had to predict where it resided. Predictions did not necessarily reduce uncertainty, nor did they constitute simple calculations based upon technical data.[18] Temporal predictions constitute one of the core mandates of modern science, but the marriage of science and prediction has been a relatively recent phenomenon that occurred in the twentieth century.[19] In the eighteenth and nineteenth centuries, earth scientists avoided making predictions and concerned themselves foremost with observing geological phenomena and using that evidence to explain earth processes.[20] Oil's hiddenness *required* prospectors to predict, increased the uncertainty of their explanations, and underscored the interdependence of and complex mixture between scientific, social, and political factors at the center of any prediction.[21] Those who vied for the authority to determine where this resource existed often held very different stakes in the matter. Predictions of where oil resided often reflected epistemological contests and highly politicized struggles for profits, power, and authority.

CHARISMATICS

Even as science and technology advanced, the necessity to predict oil's location prompted vernacular prospectors as well as professional geologists to attribute their successes to luck. Especially in untapped regions, the abundance of oil residing at or just below the surface increased the probability of success, and the more holes drilled, the more they understood underlying strata.[22] O. W. Killam, who earned a reputation as a doodlebug in Laredo, Texas, illustrated the necessity of luck by conceding that a doodlebug instrument "makes you spend your money drilling holes in the ground," but it was not so much the machine as "the law of averages [that] will finally hit you a pool of oil."[23] Modern-day geologists who employ sophisticated technology belie the notion that finding oil always requires significant uncertainty when they repeat a sentiment long familiar to oil industry insiders that "oil is where you find it."

People who attributed successful oil finds to luck expressed a worldview ordered by chance and fortune. As cultural historian Jackson Lears argues, throughout the nineteenth and into the twentieth centuries, American culture reflected an ongoing and persistent tension between cultural values of chance and control.[24] People who embraced luck equated moral worth with uncertainty, contingency, and grace rather than with a person's net worth.[25] Industrial capitalism partially eroded the culture of chance that had been Americans' dominant idiom and replaced it with a culture of control that enshrined managerial thought and emphasized efficiency at the expense of morality.[26] Even into the twentieth century, however, as petroleum geologists began to rationalize and codify prospecting, Everette L. DeGolyer, one of the most influential geologists of his time, noted that "it takes luck to find oil." At the heart of this emerging new science lay a combination of luck and skill, "but don't ask what the proportion should be. In case of doubt, weigh mine with luck."[27]

Luck functioned as a historical construct that reflected American values over time and place, and the newer the settlement, the more people invoked luck's power. Particularly in frontier regions, where values remained in flux and freer from challenge by mainstream orthodoxy, confidence men who embraced an ethic of fortune stepped into a void and gave human form to the culture of chance. One such setting was the Mississippi riverboat Herman Melville used for his 1847 novel *The Confidence Man* in which a beguiler adopts a series of disguises in order to defraud trusting passengers as well as other schemers. Whether referred to as sharpers, grafters, or cheats, these characters abounded in nineteenth-century America.[28] Fluid social conditions, values, and mores on the edge of Euro-American settlement lent themselves to the machinations of these practitioners of deception.[29] Confidence men flocked to mining towns to orchestrate ruses that would generate profit. Some confidence men salted mines by planting mineral deposits to convince speculators to invest in developing the project and then disappeared in the dark of night.[30] Their profits depended upon people willing to take a risk, speculators wedded to the culture of chance.

Like mining towns, oil boomtowns attracted confidence men who lured speculators willing to embrace chance. Fraudulent oil promoters emerged soon after the first boom in Pennsylvania in the 1860s, leading a music publisher to produce a songbook, *Oil on the Brain Songster*, satirizing their tactics. E. Pluribus Oilum, the author of a song titled

"Famous Oil Firms," wrote lyrics poking fun at the names of these illegitimate companies:

There's "Ketchum and Cheatum,"
And "Lure 'em and Beatum,"
And "Swindleum" all in a row;
Then "Coax 'em and Lead 'em,"
And "Leech 'em and Bleed 'em,"
And "Guzzle 'em, Sink 'em, and Co."[31]

Tricksters often created companies and adopted names to suggest affiliation with a successful well or company, such as the "Rockefeller Oil Company of Beaumont." Just as confidence men salted mines, so too they salted oil wells. In one case a promoter sold a productive well to investors who later discovered that a pump merely recirculated the oil they observed flowing from the ground.[32] These practices and others like them grew so notorious that "the oil promoter" emerged as a stock folk character recognizable in most oil towns.

Some oil prospectors devised doodlebug machines solely for the purpose of defrauding investors. One example consisted of bells, whistles, and dials attached to a large black box that a man operated while seated and covered with a shroud while four men carried him and the device over prospective oil fields. When the bells rang, the operator declared that oil lay below. Allegedly the shroud prevented discovery of the machine's secrets, but it may have also concealed the operator inside who rang the bell.[33] Doodlebug prospectors perpetrating fraud intended the elaborate appearance of their devices to deceive investors and kept the "technology" on which they operated a mystery in order to perpetuate the ruse.

To minimize fraud and opportunities for confidence men to manipulate people's belief in luck, vernacular prospectors fashioned a range of methods and techniques for finding oil. As long as the principles of petroleum geology and engineering remained undeveloped, the knowledge that people generated above ground remained contested terrain and told them very little about oil's subterranean habitat. The range of methods vernacular prospectors devised generally fell under the label creekology. Literally the term referred to the supposed relationship between the flow of a creek and the presence of oil, but it grew to encompass many techniques and acquired a generic meaning that applied to any nonscientific approach to prospecting.

Social theorist Max Weber considered people who exhibited or professed supernatural abilities throughout history to be powerful leaders in their societies specifically because they derive authority in a manner fundamentally different from bureaucratic officeholders or professionals with expertise and training. Weber had much to say about the characteristics of a modern bureaucracy, but he primarily defined it as a jurisdiction overseen by "professionals" who have undergone a prescribed course of training that qualifies them to administer its rules, laws, and regulations.[34] When societies require leadership in matters that transcend daily economic routines particularly "in moments of distress," they look to leaders with "charismatic" qualities for guidance. Charismatic leaders possess "supernatural" abilities and "specific gifts of body and mind" not accessible to others. Unlike professionals in bureaucracies, charismatic leaders acquire authority by demonstrating their gifts to adherents. Magicians and shamans possess such authority but so too do charismatic doctors, judges, military leaders, and hunters. Bearers of charisma determine the limits of their power by proclaiming their tasks and demanding that others follow their lead. A prophet's authority collapses when he fails to engender belief in adherents and they no longer feel compelled to follow.[35] Not everyone in the oil field endorsed charismatic prospectors' authority or believed in their methods, but the latter's presence left enough of an impression on people who witnessed them to leave testimonies of what witnesses had seen.

Although some prospectors who claimed charismatic power functioned as confidence men, others held a deep conviction in the knowledge they proffered. Charismatic prospectors surfaced in most oil regions from Pennsylvania to California and claimed a unique ability to locate oil. Some oil finders who claimed supernatural powers had unscrupulous intentions to defraud the public by charging a consulting fee in exchange for guidance about where to find oil. However, other oil finders unabashedly embraced their supernatural abilities and alleged powers as did the individuals who employed them.[36] One visitor to the oil region noted the increasing presence of people "claiming to be gifted with extraordinary powers."[37] Shortly after the discovery of Drake's well, another contemporary observed that "a new class of people has sprung into existence under the cognomen of 'oil smellers,' who profess to be able to ascertain the proper spot by smelling the earth."[38] Some within the industry remained skeptical of these powers, but others could not refrain from consulting those claiming powers when faced with no alternatives. One contemporary observed that clairvoyants exerted "more real power among the operators than the latter are willing to openly concede."[39] Oil companies typically did not mention in their prospectuses whether they had hired a consultant that investors might consider unorthodox, but they justified the expenditure because the small fee amounted to "an inconsiderable item in the general expense, seeing we mean to bore any how."[40]

Another type of vernacular prospector, the oil seer, claimed to possess an ability to locate oil without having to observe the landscape. An Italian woman named Augusta Del Pio Luogo felt "little shocks passing from her feet to her head, causing distinct pain" when she walked through a field containing oil or water.[41] Luogo first recognized "her powers" as a child and her successes proved so remarkable that oil companies began hiring her. She sometimes used instruments like divining rods but made no attempt to correlate specific identifiable land forms or structures with underground reserves. Her body provided the vehicle for detecting oil. Similarly, when the seer Evelyn Penrose walked over an oil field she felt a "violent stab in the soles of my feet like a red-hot knife."[42] Any indication of where oil reserves might lay rested solely on the seer's advice and bore no relationship to the physical terrain. Many other seers did not even require close proximity to a potential site in order to locate oil.[43] Ruth Bryan (also known as Madame Virginia) of Abilene, Texas, claimed to determine whether a farm contained oil simply by talking to its owner.[44] The environment remained irrelevant to her approach: "I can tell them how the land looks before I ever get on the land." In one instance, she intuitively sensed where oil lay when she passed over it while riding a train at night.[45] Not having to identify specific features on the landscape allowed Madame Virginia to retain a high degree of authority among oil investors. She enhanced her authority further by not identifying the source of her alleged ability.

Surveying the landscape for specific geological or geographical features played no part in the methodology employed by Guy Finley, also known as the x-ray-eyed boy. Finley received visions of oil beneath the ground at a very early age. He retained this alleged ability into adulthood and earned a reputation among investors eager to profit from oil. One group of investors put Finley's abilities to a test by burying two barrels, one filled with water and another with oil.[46] Finley claimed to have located both barrels successfully. To maximize his effectiveness, though, Finley preferred to search after the sun had gone down — the darker the better. His preference for working in the dark of night may have reflected his desire to perpetuate mystery or shield onlookers from some kind of subterfuge. By keeping oil investors guessing about their supposed powers, vernacular oil prospectors who claimed supernatural abilities carefully protected the authority optimistic investors eagerly bestowed upon them.

Charismatic prospectors did not consistently succeed at finding oil, but they proved adaptable and devised new methods when their authority waned. When clairvoyance and dreams failed to produce oil or convince investors to part with their money, some prospectors began to employ a variety of oil-finding devices such as divining rods. The divining rod and other devices with a more sophisticated appearance acquired the name of doodlebug, which applied to either the apparatus itself or the person using it. Doodlebug devices came in a variety of shapes and designs, but the divining rod served as the most recognizable example. Also known as dowsing or witching, prospectors throughout history have used divining rods, or wiggle sticks, for a variety of reasons that included locating minerals, water, buried treasure, and even escaped criminals.[47] The instrument usually consisted of a Y-shaped tree branch, often from a peach tree but sometimes a hazel or willow.[48] One oil prospector, Jonathan Watson, resorted to divining rods only after the advice of spiritualists began to fail, but the witching-stick operators he hired proved no more successful with each well they located coming in dry; these failures tempered Watson's faith in the device.[49] Every failed attempt at locating oil with a divining rod produced doubt in the minds of some contemporaries as to the credibility of the device and its operator.

Scientists frequently rejected outright the efficacy of divining rods and either considered them instruments to defraud the public or attributed their role in successful oil discoveries to chance.[50] Yale University professor Benjamin Silliman declared that "the pretensions of diviners are worthless. The art of finding fountains of minerals by a peculiar twig is a cheat upon those who practice it, an offense to reason and common sense, an art abhorrent to the laws of nature, and deserving universal reprobation."[51] Silliman's critique revealed that he defined common sense in a manner differently than divining rod operators. Common sense represented more than just a mere factual apprehension of reality but was a historically constructed cultural system and subject to different standards of judgment.[52] An individual culture may construct a unique commonsense tradition that leads its people to accept witchcraft or magic as acceptable explanations for causation.[53] Western science profoundly shaped people's views of common sense.[54] Silliman adopted this latter perspective, but charismatic prospectors retained their own views of what constituted common sense.

By devising more elaborate devices than divining rods and declaring a unique ability to operate them, doodlebug prospectors sought to bolster their authority against accusations leveled by scientists such as Silliman. As the oil frontier moved west and into the twentieth century, more elaborate and exotic devices proliferated alongside divining rods.[55] For example, a Houston, Texas, doodlebug operator named Dr. P. S. Griffith possessed a black box that contained an oil-finding device. Tubes extended from the box, two that served as handles and a third that terminated with a plate fitting onto the roof of his mouth. He possessed a dozen metal capsules approximately three inches in length and the diameter of a pencil with labels such as oil, gas, gold, and silver. He screwed the capsule, or lug, labeled oil onto the machine and began surveying the landscape.[56] "Sometimes he would start and tremble and you'd see this lug draw down toward the ground. There he'd make a mark."[57] Griffith offered a vague scientific rationale to explain how the device worked. The doctor explained to one observer that "it was something in these lugs plus a magnetism of his — in his body — some companion property of chemistry that made it possible for him to locate these things."[58] Although doodlebugs could not prove their devices operated on scientific principles, the inability to disprove their claims prolonged the credibility they enjoyed.[59]

A doodlebug who could not describe how his device worked, however, was not necessarily trying to swindle the public and sometimes possessed a legitimate knack for finding oil. Killam could not even characterize a doodlebug device much less explain how it worked. He described it as "a little instrument that goes up and down and around and around" but grew exasperated in his explanation and confessed "there's just no way to describe them."[60] There existed "dozens" of different types and they all "work on the same principle."[61] Although Killam failed to state that principle explicitly, his description revealed that success at finding oil had more to do with the operator than with the device itself.

The body serves as an important vehicle for humans to perceive the world because it functions as a receptor that participates in the cultivation of tacit knowledge. Chemist and philosopher of science

Michael Polanyi has argued that all knowledge is personal and tacit in character and that "our body is the ultimate instrument of all our external knowledge."[62] He derived this argument from the field of Gestalt psychology, which showed how people possessed knowledge but remained unable to identify particular aspects of what they knew.[63] For example, a witness at a crime scene can recognize a perpetrator's face from among a million possibilities but cannot articulate why this is so. Polanyi argued that this Gestalt phenomenon operated both psychologically and physically, and he cast both reactions as "the outcome of an active shaping experience performed in the pursuit of knowledge."[64] Our bodies experience the Gestalt because they function as receptors that help us navigate the world, as when a blind man deciphers his path based upon the sensation it transmits to his hand through a cane.[65] Furthermore, physiologists have demonstrated that our perception of an object can be influenced by reactions within our bodies we never feel, such as muscular twitches responding to subliminal stimuli.[66] Based upon this reasoning, Polanyi concludes that tacit knowing operates "on an internal action that we are quite incapable of controlling or even feeling" and that "by elucidating the way our bodily processes participate in our perceptions we will throw light on the bodily roots of all thought."[67]

Like a blind man navigating the terrain with a cane, the most successful doodlebug prospectors also surveyed the landscape, and this activity cultivated within them an instinct for recognizing changes in topography and vegetation that indicated the presence of oil. In order to operate a doodlebug, Killam explained that "you've got to have a lot of common sense and some knowledge of oil to get any effective results."[68] His statement revealed a historical construction of common sense, an interpretation based upon the immediacy of his experiences looking for oil.[69] His emphasis of personal knowledge rather than the machine's proficiency reinforces Polanyi's point about the "bodily roots of all thought." Although an avowed believer in doodlebug devices, Killam considered the individual's common sense and knowledge of how oil accumulated the most important ingredients to finding oil. He could not articulate exactly what role the device played in a successful find, subscribing to the "theory that if you knew what to look for you could see an oil field on top of the ground."[70] Instinct provided the prospector's most effective attribute "cause if you'll go to any oil field you'll see that it differs a little bit from the surrounding territory."[71] Successful prospectors cultivated an intuition or knack for finding oil by surveying the environment and the doodlebug device may have facilitated that process.

The case of Johannes Walther makes the connection between a doodlebug apparatus and its operator's body more explicit. Humans' bodies and the changes they underwent offer telling clues about how people settled and took possession of the land during the course of westward expansion. The ailments that afflicted people and the remedies they imbibed offer testimony to the struggles of inhabiting and taking possession of mountains, pastures, streams, and swamps.[72] Similarly oil prospectors' bodies serve as palimpsests that reveal landscapes' imprint. Walther observed this dialectic when conducting an experiment in which he observed two groups of prospectors operate divining rods. Members of the first group had experience operating the device, but the second group, a control, consisted of people with no divining rod experience. Walther concluded that many people have the ability "to feel the presence" of oil and other substances but to varying degrees. Their ability to sense oil with the divining rod resembled the manner in which musicians display sensitivity to sound through their instruments.[73] An encounter with oil "stimulates in some way" the operator's nervous system "so that a certain feeling is caused in their tissues," and their muscles react whether they held the device or not.[74] When a divining rod moved, Walther explained, it merely reflected the operator's muscular reflex. The sincerity with which divining rod operators believed in their methods particularly struck Walther. A geologist who reviewed his findings expressed skepticism, arguing that the conclusions seemed "entirely improbable in the ordinary sense of the word."[75] Only if tested "under correct scientific control" could he accept that "wigglestickmen are not frauds, whether consciously or unconsciously so."[76] Like Silliman, this scientist discounted knowledge diving rod operators cultivated because of its tacit and local character and vested more authority in "correct scientific" knowledge determined by him and other trained professionals.

Doodlebug devices, which geologists dismissed as unscientific or fraudulent, held symbolic meaning for vernacular prospectors but in much the same way as ritual artifacts held practical benefit. Symbolic processes such as divination and performance of magical rites offered therapeutic relief and in some cases extraordinary physical strength or recovery from organic disease. Material artifacts play a role in rituals. Many cultures invest apparently trivial objects with a powerful force anthropologists call "mana" that pervades, sustains, and rules their spiritual worlds. Examples of sacred artifacts include cowrie shells, palm nuts, smooth stones, soothsayers' bones, dice, coins, and a lucky rabbit's foot. The trickster plays a central role in these cultures, and chance figures prominently in people's lives. Only by abandoning a linear, progressive understanding of the world do cultures of chance and the ceremonies at their core make sense. Divination and magical rites function as performative acts, not prelogical or prescientific behavior. In the oil industry diving rods and other forms of doodlebugs reflected people's recognition that chance mattered greatly in their potential opportunities. Scientists who compared prospectors' use of divining rods to scientific oil explorations committed what philosophers call a category mistake.[77]

PRACTICAL OIL MEN

Another category of vernacular prospectors who generally avoided divinatory rituals but who also remained apart from professional geologists was practical oil men. The very word "practical," which people used to characterize these men, attests to their temperament, as well as the hands-on, commonsense knowledge they used to great effect.

While at times many practical oil men relied upon the techniques of doodlebugs, they differentiated themselves by relying foremost on systematic surveying and observation of the landscape before expending the capital to drill. Long before geologists articulated and applied geological principles to guide prospecting, practical oil men constituted key figures in the oil industry because of their innovations in exploration and production techniques.[78] These prospectors also referred to themselves as wildcatters, another term that described their independent mentality and spirit. Rather than follow the horde of promoters who believed drilling next to a successful well maximized their chance of success, an oil man who knew how to read the landscape might drill in a place that no one else even considered — out among the wildcats. These independent entrepreneurs formed the bedrock of the industry and found much of the nation's oil well into the twentieth century.

Historian Frederick Jackson Turner would have recognized many of the traits the men who settled America's oil frontier used to describe themselves. According to Turner, frontier settlers possessed a "practical, inventive turn of mind," and from their experiences emerged the "dominant individualism" that characterized American character as well as the "coarseness and strength" that accompany "freedom."[79] Turner mentioned the oil industry occasionally in his writing. In places like Oklahoma, he wrote, Indian Territory "passed away" and cities arose, and "it was not long before gushing oil wells made a new era of sudden wealth."[80] By the twentieth century large-scale production and the trend to amass capital "found in America exceptional freedom of action."[81] At the same time, "the old pioneer individualism is disappearing," and the self-made man became "the oil king."[82] Indeed, many practical men who succeeded on America's oil frontier amassed capital equivalent to kings, but probably more did not succeed to this degree or even find a drop of oil. Even for the less successful oil men, however, Turner's triumphal narrative resonated powerfully because they identified strongly with the traits he ascribed to Americans such as independent, individualistic, self-reliant, and democratic. Whether historians agree with Turner's frontier thesis or not, oil men embraced his version of westward expansion as characteristic of their experiences. In doing so, practical men cast themselves within a heroic context of American exceptionalism. Their ingenuity and inventiveness contributed greatly to the role of the United States as the leader in world oil production prior to World War II.

Prospectors located much of the oil that catapulted the United States to world dominance in oil production with methods professional geologists discredited as nonscientific. Field work provided the key to practical men's success because this activity enabled them to cultivate a form of geological reasoning based upon correlations that they established between surface topography and oil beneath the ground. Although not recognized by the scientific community as professional geologists, practical oil men in the Petrolia, Pennsylvania, field were "nothing if not geological. . . . Nearly every operator is ready to discourse learnedly on rocks, formations, strata, shales, sandstones."[83] While traveling the countryside, they "engaged in a kind of blind man's bluff with nature" in order to decide whether or not to lease a particular tract of land.[84] Any decision to lease involved a significant degree of risk without the visible presence of oil, but "the more capable prospecting was guided by a combination of instinct, experience and rule-of-thumb geology."[85] Thus drilling for oil on the basis of a hunch did not necessarily translate into guesswork. Rather, their interactions with the natural environment led them to decipher potential correlations with land formations above the ground and oil that lay beneath. That some prospectors were consistently more successful than others suggests that field experience developed within them a stronger instinct for identifying surface structures that might lead them to oil. In this way their rule-of-thumb geology grew out of visual and physical encounters with the environment.[86]

Field work so significantly influenced prospectors' instincts about where to drill that even when they approved of a location on what they thought was superstition, the decision often possessed a geological rationale. No clear line demarcated the methods of vernacular prospectors, and sometimes a practical man employed methods typically charismatic and vice-versa. This overlap in methods was apparent in one prospector's choice for the best site to drill. Field work led him to "a decision that oil was to be found most probably in association with the base of some mountain ridge or creek bed or some other geological feature of the surface."[87] Although his reasoning lacked a sound geological rationale, the idea that a relationship existed between surface formations and oil beneath the ground possessed some degree of logic. The same man easily dismissed such logic, however, and allowed superstition to inform his decision by concluding that "the place that seemed most attractive to me had for many years been used as a burying ground. It was a churchyard cemetery."[88] Some prospectors so vehemently subscribed to the notion that oil resided beneath cemeteries that "when oil was found in some neighborhood, any graveyard there was soon encompassed by a forest of derricks."[89] The superstition took such a hold that other oil men rapidly began leasing land in nearby graveyards.[90] On one hand, the contention that oil always resided beneath graveyards places this approach to oil finding within the realm of superstition. On the other hand, a more rational explanation for drilling in cemeteries existed even if prospectors only recognized it intuitively.

Cemeteries often yielded oil not for mystical or supernatural reasons but because the hills they sat atop consisted of geological structures that served as repositories for oil. While drilling in a graveyard did not guarantee success, prospectors often found oil beneath them because structures such as anticlines and salt domes provided high ground that served as ideal burial sites. In fact, most graveyards situated in oil regions contained oil.[91] The one thousand barrels one man extracted from a cemetery led him to reflect, "I think back on that churchyard drilling as one of the best of my early operations."[92] Although a supernatural belief that the Lord blessed a churchyard cemetery's holy ground by bestowing it with oil motivated some men to drill there, the decision more often resulted because businessmen lacked a systematic method for finding oil and followed a tradition that had proven successful in the past.[93] The notion grew so popular that it evolved into a folk tradition, prompting some congregations to lease out their churchyard cemeteries.[94] Similarly, just as cemeteries yielded oil because of their elevation, sawmills indicated sites to avoid because builders typically situated them on lower ground or in geological structures known as synclines where oil tended not to accumulate.[95] Thus sound geological principles explained the existence or absence of oil even when superstitious or supernatural explanations appeared more obvious.

Creeks as possible drilling sites also attracted vernacular prospectors, who often found oil nearby, but success resulted from underlying geological formations rather than the mere presence of water. Prospectors grew so convinced that oil yielded by a single well bore a relationship to the twists and turns of a creek that they used the term "creekology" to describe practitioners who studied a creek's flow. The presence of oil had less to do with a creek, however, and more to do with the region's geology. The downward slope of anticlines often formed valleys through which creeks, streams, and rivers flowed. Drilling next to the water met with success because the anticlinal structure trapped oil inside, not because of oil's relationship to the creek.[96] In this case oil men unknowingly applied geology even when they subscribed to the folk tradition that they could find oil by drilling next to creeks. Prior to 1885 petroleum geologists had not consistently and unanimously articulated principles that prospectors could rely upon as tools for exploration.[97] Nevertheless, oil men began to realize that actively studying the composition of the earth could lead them to oil.[98]

One of the earliest attempts to formulate a systematic method for finding structures that revealed the presence of oil resulted in the mistaken idea that large underground crevices or fissures provided cavernous reservoirs for oil.[99] This idea constituted oil country folklore that remained current during the first twenty years of the industry and probably originated with Uncle Billy Smith, who drilled Drake's well. Smith vowed that when the hole reached sixty-nine feet, he encountered an opening that caused his tools to drop six inches.[100] The idea so captured the imagination of drillers throughout the Pennsylvania oil region that "it was the popular belief that a fissure must be struck in the oil sand or a well would be a failure."[101] The idea that the quantity of oil bore a direct relationship to the size of an underground crevice possessed no geological validity, but drillers adopted this misconception as truth. If a driller happened to locate a productive well, he suddenly "recalled" that indeed his drill had dropped at a certain point. Subsequent production of the well influenced his recollection of how far the drill had dropped.[102] The more productive the well, the greater the distance the driller imagined his drill had dropped in order to reflect the size of the supposed crevice. The manner in which practical oil men transformed the geologically mistaken notion of crevices into truth provides another example of how their bodies functioned as receptors that perceived tacit knowledge of the subterranean world. In this case the explanation of what they perceived proved erroneous.

The belt-line theory of oil accumulation was also a folk tradition although different from drilling next to creeks or into large underground crevices because it possessed a more geologically sound rationale and thus provided a fairly reliable prospecting method. Vernacular prospectors correctly reasoned that oil often lay in patterns beneath the surface. A practical oil man without any geological training, Cyrus D. Angell observed in 1861 that a number of successful wells tended to occur in a pattern irrespective of flowing water.[103] Based upon his experiments, Angell concluded that oil lay in continuous belts that ran along a straight line in a northeast-southwest direction and at an angle of 22½ degrees longitude.[104] (The angle of the belt varied depending on the geology of a given location.) Following this reasoning, Angell could locate a belt by plotting its course from the surface and determine that creeks bore no relationship to the linear belt of oil.[105] He successfully demonstrated this belt-line theory in 1871 and again in

1873 by locating productive wells after correlating the stratigraphy of potential sites with that of known producers.[106] Even though some of Angell's assumptions and principles were exaggerated, invalid, or parochial in the manner he applied them, he provided a method of exploration based upon quasi-geological principles in order to eliminate the role of chance.[107]

Despite limitations of the belt-line theory, it grew in popularity and displaced other less geologically sound methods. The belt-line theory changed how prospectors conceived of where to find oil. Previously, they remained so blindly attached to the idea that oil resided along creeks they rarely drilled at high elevations that looked down into the valleys where water flowed. Belt-line theory prompted them to think differently. Prospectors began to reason that regardless of where the creek flowed, if two wells "are alike in depth, and appearance of oil, and of the rocks bored through, I should be inclined to think they are all on one belt."[108] This new way of thinking placed a premium on gathering geological data: "In nearly all the shanties, or in the engine houses adjoining the wells, or else in the offices of the owners of the wells, were preserved specimens of the different kinds of rock found in each well. They asked for little specimens of these to compare with similar ones from all other wells."[109] The belt-line theory widened oil prospectors' perspectives from the confines of a single river valley or creek bed to a much larger region. Gathering and correlating geological specimens led them to see that the geology of two locations "six miles apart, separated by a mountain, were almost exactly alike."[110] A broader perspective paid dividends because identifying relationships between two distant sites meant that "they could find good wells all along the line," or belt, that separated two producing wells.[111] Not all practical oil men, however, consistently applied this new way of thinking.

Geological reasoning was key to the success of the belt-line theory, but some prospectors elaborated upon it with their own uninformed hunches and poor intuition and thereby diminished its value. Many practical oil men ignored the geological criteria Angell outlined in his hypothesis and formulated belt-lines based on their own idiosyncratic philosophies.[112] Whereas Angell designated the oil line's longitude at 22½ degrees based upon the geology of the region he examined, among practical men additional degree lines grew in popularity without regard to the unique geology of a particular locale. Drilling for oil in a pattern, or along a belt-line, at the surface without taking surrounding geological structures into consideration meant failure before drilling even began.[113] Prospectors who applied Angell's theory in their own idiosyncratic manner undermined the quasi-geological thinking that undergirded the theory and proffered in its place a speculative conception of locating oil based more upon hunch and intuition than observable facts. Stripped of its geological principles, the belt-line theory held less benefit. Nevertheless, it remained the most popular approach among practical men until about 1920.[114]

GEOLOGISTS

Geologists who could rationally explain to investors why one site recommended itself over another provided utilitarian and potentially valuable information, but the science this emerging class of professionals offered represented a defensive challenge to the knowledge vernacular prospectors possessed. As previously mentioned, common sense refers to more than just a matter-of-fact apprehension of reality. Anthropologist Clifford Geertz contended that people acquire commonsense knowledge through interactions with local environments by which they cultivate a colloquial wisdom enabling them to deal effectively with everyday problems by judging, assessing, and apprehending their surroundings.[115] However, some people do not always perceive the world judiciously, intelligently, perceptively, or reflectively.[116] What one interest group considers readily apparent may be highly contested by another. Common sense, like other cultural systems such as science, religion, and art, is historically defined and can vary dramatically over time and space and from one culture to another. Because of its variability, determining what qualifies as common sense can be a highly contentious exercise. Therefore, the legitimacy of common sense rests upon the conviction of its value and validity among those who possess it.[117]

Conceiving of common sense as a culturally and historically constructed body of knowledge helps to illuminate why geologists dismissed both charismatics and practical men with accusations of superstition and a naive belief in magic and witchcraft. The cry of witchcraft reflects less a questioning of how the world is put together on the basis of religious, philosophical, scientific, or moral principles and more of an attempt by a cultural group to "block such questions from view."[118] Rather than accepting scientists' allegations of witchcraft as a form of primitive metaphysics, we should consider the efforts of vernacular prospectors as efforts to certify the seen world that they had witnessed and not as celebration of an unseen mystical world.[119]

In this context, geologists who dismissed the methods of vernacular prospectors outright accounted for their own inability to articulate clearly and systematically how to find oil. To accrue authority as professionals who offered utilitarian value to the oil industry, geologists faced the task of explaining how investors could minimize financial risk by drawing upon identifiable knowledge to increase the chance of finding oil. This knowledge remained highly rudimentary in the nineteenth and early twentieth centuries. In fact, it is still unfolding. As Geertz put it, "Men plug the dikes of their most needed beliefs with whatever mud they can find."[120] Accusing vernacular prospectors of quackery, mysticism, or superstition served as the mud geologists used to plug the cracks in a body of knowledge that they had only begun to identify, articulate, and delineate.

Geologists' lack of patience for vernacular prospecting intensified into the twentieth century, in part because their authority remained tenuous and vernacular methods — divining rods in particular — still appealed to many people. In 1917 the United States Geological Survey published a history of the divining rod in response to a large number of inquiries it received annually regarding the efficacy of the device.

The report's author acknowledged that divining rods possessed an appeal throughout time and in different countries but saw no utility in them. He regarded them "a curious superstition" but one that "still has a strong hold on the popular imagination." He warned that "uneducated persons" often fell prey to tricksters who attributed "the working principle of such a device [to] some newly discovered and vaguely understood phenomenon."[121] Even at this late date, geologists refused to consider the possibility that the mystery surrounding the principles upon which doodlebug devices operated might be explained by understanding the commonsense knowledge of the prospectors operating them.

A variety of different practitioners cultivated local knowledge through their work in Pennsylvania and other states as the oil frontier moved westward after 1859 and into the twentieth century. Geological explanations of where oil accumulated remained rudimentary, disorganized, and contested throughout this time frame. A lack of consensus on the geological principles that explained oil accumulation provided opportunities for a variety of practitioners to claim authority and power for knowing nature and its relationship to oil. Economics alone cannot explain all prospectors' motivations because culture also accounted for their actions. Opportunities abounded for people with different beliefs and practices to claim nature as their own.[122] People who cultivated local knowledge experienced nature directly and intimately through their bodies and through their work. Finding oil in this manner could be a deeply personal experience for some, resulting in spiritual revelation, intense pain, or simply an opportunity to demonstrate a version of common sense. While the principles of petroleum geology remained vague and ill defined, authority was up for grabs.

2

Collaborative Authority

Petroleum geology emerged over the course of the nineteenth century as a contested practice in which different constituencies formulated knowledge by fashioning relationships to nature through their physical and intellectual work. Many histories of science relate a trajectory in which a loose assortment of elite practitioners commonly labeled gentlemen scientists began to formalize their activities early in the nineteenth century and to form professional associations; by century's end they had fashioned an efficient and systematic body of knowledge private industry could comprehend and utilize.[1]

The science of petroleum geology was never so clear and stable as this narrative presupposes. A story line that relates how scientists gradually emerged to codify geological processes of oil accumulation dismisses the strong cultural component of laymen whose experiences in local environments informed the process of knowledge-making.[2] Practical men prospected for oil and performed the vast majority of physical labor involved in drilling, which provided them with numerous opportunities to observe the color, odor, texture, and consistency of the oil-laced soil their efforts uncovered at well sites. Experiences in nature invested them with authority to deduce from their observations when to stop drilling, when to drill deeper, or where to drill next. Disputes over the meaning of this knowledge intensified when scientists began examining well samples in an effort to derive theories that carried translocal and potentially universal explanatory power. Scientists who built upon these data to articulate geological theories acquired a new kind of authority by the 1880s through their capacity to generate knowledge that was codified, systematic, and universal but most importantly that explained how and where oil accumulated.[3] Professional scientists acquired this authority in part from their intellectual efforts but also from their ability to capitalize on the knowledge practical oil men generated from working in nature.

Geologists who acquired authority did not operate in unanimity or consensus but engaged in epistemic contests with one another, further underscoring the contested nature of relationships that all practitioners forged, professional and laymen alike, between the environment and the ideas and practices of finding oil. Knowledge about petroleum existed as theories in the minds of geologists as much as it took form in the physical labor and sensory experiences of practical men.[4] Casting geologists as theoretical and laymen as practical overemphasizes differences between the two and obscures how *both* groups theorized and physically labored to find oil. Still, these categories, although overstated, help to convey that the locus of authority among oil prospectors shifted from those whose work practices were local, experiential, and tradition laden to professional geologists who strove to fashion universal and systematic ways of knowing the environment.[5] This shift occurred neither naturally nor inevitably and involved much discussion, dispute, and disagreement among geologists. Like practical men, geologists fashioned relationships to nature in highly individualistic ways, and their experiences in nature shaped the geological theories they formulated. Diverse environments in which oil was situated shaped their perceptions and often exacerbated disagreements among them. Thus geologists contested the knowledge of oil accumulation among themselves as much as they disputed the ideas and methods of practical oil men.

What all these contests for authority reveal and what this chapter explores is the idea that the science of petroleum geology was fundamentally a collaborative effort between different constituencies shaped by the environment's surface and subsurface geological makeup. No single constituency or person won the turf battles fought over oil's location because science, and petroleum geology in particular, grew as the result of a long and ongoing collaboration in which various practitioners participated even if not always harmoniously. Emphasizing how an array of geologists and practical men collectively shaped the field of petroleum geology reveals what one historian has called a "complex web of social and cognitive interactions" that bound a "network of colleagues, in collaboration or rivalry or both."[6]

In the oil industry collaboration and rivalry took several forms. Some geologists recognized the difficulty of amassing and interpreting enormous amounts of data and collaborated more easily with each other to meet their employers' demands. Others proved more intransigent, challenging the theories of their peers or taking from nonexperts knowledge that they considered crudely factual and that required a geologist's purview to make sense. The best geologists struck a tenuous balance between a wide assortment of facts gained through local experiences and broader theories that although more abstract and arcane, possessed significant explanatory power. Geologists did not always agree that a particular theory followed from a selection of facts, but their disputes helped to advance geological knowledge and precipitated the field of petroleum geology.

NATURE'S AGENCY AND VISUAL CULTURE

Throughout the nineteenth century, nature more than any other factor confounded prospectors' efforts to find oil. As oil fever spread in Pennsylvania, prospectors and scientists surveyed the landscape's surface topography and studied the region's geology. Nature foiled their efforts, especially when oil accumulated because of geological processes too deep for the searchers to observe. Although prospectors and scientists increased their fund of knowledge through surface reconnaissance, they remained ignorant of the intricacies of Pennsylvania's subsurface geology. Even when geologists successfully solved the puzzle that explained how geology trapped oil in one locale, a different environment presented them with challenges to formulate new theories and articulate verifiable principles for prospecting elsewhere. Practical men who employed the belt-line theory met with mixed success, and when this theory did not lead to oil, they contended that fortune had not shined upon them. Geologists saw prospecting differently and felt "it is a mistake to suppose that all oil discoveries are entirely the result of chance."[7] Rather than submit to chance, geologists adopted the mission of formulating knowledge that would enable them to begin understanding how nature distributed oil in its various environments. Professional scientists attempted to rationalize and codify explanations for oil's location by studying the environment's geological makeup. One engineer observed of Pennsylvania's oil wealth that "nature had a system in this, as in all her other works."[8] Explaining how that system functioned proved no easy task.

Maps, diagrams, and other visual images provided the means by which geologists began to explain how nature's system functioned and enabled them to capture nature's agency by depicting visually the multiple and varied relationships between geological processes and oil accumulation. Geologists wrestled with the task of demonstrating how facts they gathered conducting field work substantiated their geological theories. Images that directors of state geological surveys included in their reports reflected a visual language that was part of a developing self-conscious new science practitioners called geology.[9] The necessity of relying upon three- and four-dimensional imaging to understand subsurface geology has become so ingrained in modern geological and geophysical explorations for oil that scientists and lay people alike take for granted the use and availability of visual images to peer beneath the earth's surface. Visual images, however, were not always used to study the earth. Beginning in the late eighteenth century and reaching fruition by 1840, an increasing quantity of various types of illustrations — maps, cross-sections, color slides, and diagrams — appeared in books and journals and originated from diverse traditions such as mineralogy, cosmogony, natural history, and mining practice. Out of these traditions the field of geology coalesced, acquiring unique intellectual goals and institutional forms. The construction of a visual language paralleled and was part of the historical process in which geology emerged. The forms of visual expression accompanying that science complemented written descriptions and theories and communicated ideas and observations when words proved cumbersome.[10] Geological images potentially clarified abstruse theories by depicting them in visual form. Diagrams such as cross-sections that portrayed vertical dimensions of the earth represented progressively abstract and theoretical notions of subsurface geology.[11] As images grew more abstract and formalized, they acquired the potential to carry an increasing load of meaning but also to alienate lay audiences for whom they appeared too arcane. Geological surveys proliferated throughout the nineteenth century, and images that had become integral to geology's visual language potentially facilitated survey directors' primary goal of generating knowledge that lay people could apply to locate and profit from natural resources.

Geological surveys generated information that potentially resolved or perpetuated scientists' theoretical disputes and disseminated practical information to the taxpayers who funded them. State legislatures across the country began funding geological surveys in order to promote economic development based on geologists' discovery and examination of natural resources. State surveys started in the South in the 1820s, quickly spread throughout the rest of the country, and within two decades had grown into important institutions for geological research.[12] Legislators appropriated money to establish surveys with the intention of disseminating information to all social levels of society.[13] Particularly in the Jacksonian era, lawmakers hoped to democratize education by requiring surveys to make scientific information available to common men rather than just to an educated elite.[14] A survey's ability to generate practical information that potentially improved an individual's material well-being or enhanced the state's economy determined whether the legislature appropriated the necessary funds. Scientists and legislators sometimes disagreed, however, over what constituted practical information.

A survey director who failed to deliver information the legislature and its constituents could easily comprehend and potentially apply confronted a politically volatile situation. Some geologists pursued scientific agendas that reflected their theoretical interests and merely paid lip service to legislators' demands for utility.[15] These geologists might find themselves unemployed or their funding slashed when legislatures deemed their work lacking educational or economic value. James Hall directed the New York State Survey and contributed to Iowa and Wisconsin's surveys and provided the classic example of a geologist who subordinated his work's practical application to its scientific relevance.[16] Rather than perform the humdrum work of locating and testing building stones or other natural resources, Hall published reports with detailed descriptions and illustrations of fossil shells accompanied by expensive reproductions of steel-engraved plates.[17] Uncertain how this work might translate into economic gain, the Iowa legislature discontinued its survey, Wisconsin refused to reimburse him for costly engravings that illustrated his reports, and both legislatures objected to funds for expensive and arcane books that allegedly benefited only a small group of scientists.[18] What Hall considered practical, legislators viewed as theoretical science without apparent benefit to taxpayers. Accordingly, politically astute geologists whenever possible refrained from designating their work solely theoretical or practical in order to avoid alienating their peers within the scientific community or the legislatures who funded them.[19]

FIRST PENNSYLVANIA GEOLOGICAL SURVEY, 1836–42

Decades before the discovery of oil in Pennsylvania, businessmen and speculators who recognized the potential to profit from mining the state's anthracite coal, iron, and salt lobbied the legislature to fund a geological survey.[20] The governor appointed Henry D. Rogers state geologist. Rogers directed the state's first survey from 1836 to 1842. Rogers's contributions to the field of geology elevated him to a position of importance within the American scientific community. He had also served as New Jersey's state geologist and his brother William served as Virginia's state geologist. Given the two brothers' intellectual correspondence and the close proximity and geographical similarities of the states in which they worked, much of the knowledge each man generated built upon or elaborated what the other had discovered.[21] Throughout their collaborations, the brothers spoke frequently of the political pressure they felt from legislators to produce practical knowledge. Henry considered the Pennsylvania state legislature a "tribunal to which I have to bow" for money and complained sarcastically about one senator who refused to vote for a bill funding the survey because it neglected other forms of knowledge such as "phrenology, physiognomy, animal magnetism, and the highly important science of *water-smelling*."[22] Rogers used his position to claim authority over what constituted legitimate scientific inquiry but learned that politicians and scientists did not always see matters similarly.

As director of Pennsylvania's survey, Rogers felt confident he could generate information that met both the politicians' demands for utility and furthered his professional goals. Rogers contended that in the discipline of geology, "whose aims are eminently practical, it frequently happens that useful results connected with the arts are involved in the higher generalizations of the science."[23] To ensure that the legislature and its constituents remained content, he published annual reports that recounted the survey's activities and documented its progress.[24] He included glossaries that explained geological terms, a common technique state geologists employed to avoid alienating the public.[25] Henry followed the law's dictate to display cabinets of specimens in each county in order to familiarize miners with the local geology. Visual images in the annual reports also played a pivotal role by illustrating ideas and making large sections of text comprehensible at a glance.

Although geologists bore the ultimate responsibility for assembling and drafting these maps, the information that the geologists relied upon originated from a variety of sources outside the community of professional scientists. Pennsylvania's environment pushed geologists in the third year of the survey, 1838, to identify and map the anthracite coal region, which "had shown itself to be no child-play" because it was "complicated in structure and difficult of representation."[26] J. Peter Lesley worked as Rogers's geological assistant and possessed a particularly adept skill as a map maker, producing most of the images published in the survey's final reports. Lesley explained how studying the region took several years, and he praised Rogers's strategy of "localizing the geologists" and delegating each to gather knowledge in designated districts.[27] Geologists gathered information from the landscapes they traversed as well as from the people who inhabited them. Lesley considered this system "a great school" because it required each geologist "to struggle with all the problems of the science" and to rely upon local inhabitants "who knew where things could be looked at, but not where they could be looked for, ignorant of what the things meant, and longing to be told."[28] Lesley privileged scientific knowledge and only tacitly acknowledged that

much of his practice depended upon local residents' ways of knowing the environment. He saw lay people as ignorant but longing for the wisdom professional geologists could impart after struggling to uncover scientific secrets. Geological theories originated from the intellectual and physical labor scientists expended in the field as well as from the experiences of local residents.

2. J. Peter Lesley. Courtesy of University of Pennsylvania Archives.

Lesley solicited factual knowledge from lay people; he saw that knowledge as a tool for geologists to build theories that carried greater explanatory power. He recounted that the geological survey made great scientific and practical gains during its third year by assembling tens of thousands of facts regarding resources such as iron ore, coal, limestone, salt and oil springs.[29] Various constituencies contributed to this fund of knowledge, including Native Americans, white settlers, hunters, fishermen, miners, and farmers.[30] The information people offered proved useful, but "no one had systematically studied" these facts putting "this and that" together. Inhabitants of the districts that geologists studied merely "gossiped about at the village bar-room" and engaged in "ignorant speculation, or absurd upside down logic" about the location of resources. Without a geologist's interpretation, laymen's experiential knowledge constituted nothing more than gossip, absurdity, and ignorance. Lesley considered them ill equipped to think geologically and instead they fashioned "all sorts of geological fancies, follies and superstitions" that "befogged the minds" of people throughout the state. They remained "profoundly ignorant" of what geology had to offer them, and thus Lesley believed the survey offered a forum for edifying the masses.[31]

The survey's demise in 1842 presented an opportunity for geologists to self-reflect and consider how their practices may have contributed to its failure, but their imperiousness led them to blame laymen for an inability to understand science. Lesley particularly took no responsibility for the survey's demise. Simply put, the citizens of Pennsylvania neither "appreciated" nor "comprehended" the survey's "worth." He belied a defensive tone when explaining how the public perceived scientists in the early nineteenth century: "The language of science was then an unknown tongue, and sounded in the ears of the people like the chattering of animals or idiots."[32] When geologists disputed one another, the public interpreted their disagreements "as good evidence of the worthlessness of all their theories."[33] Geologists' chattering represented not their idiocy but their epistemic struggles to synthesize large volumes of data and reconcile conflicting interpretations among themselves. Scientists failed to win pubic favor even when they forged a consensus because "the truths in which they agreed seemed to clergy and laity alike the insanities of an exalted imagination."[34] Geology as a field of inquiry remained very much in flux by the mid-nineteenth century. Even when geologists agreed about the processes that made up the earth, religious figures and lay people felt no compulsion to consider their knowledge legitimate.

LAY KNOWLEDGE

Geologists who wallowed in self-pity at feeling unappreciated and misunderstood did not exhibit magnanimity for the laymen they derided. Lesley expressed irritation that some people who had only recently entered the mining industry claimed authority as geological experts. Recent transplants from "foreign" locales and "quite different geological regions" shed their status as "the commonest miners" only to adopt "the character and position of mining engineers in America."[35] Neither their laboring experiences nor their geological knowledge qualified them to understand the kind of knowledge the survey produced. To put a fine point on the matter, Lesley made clear his opinion of lay people and their perception of professional geologists: "Ignorant, undisciplined, obstinate, narrow minded and superstitious by nature and habit, and rendered presumptuous and dogmatic by their strange advancement, they were as unwilling to accept as they were unable to acquire a correct knowledge of our geology, so different from their own, and hated professional geologists."[36] Geologists had not failed the people of Pennsylvania whose intellects failed to grasp "a correct knowledge of our geology." Lesley's differentiation of *our* geology, which was "so different from their own," indicates his perception of a demarcation between the ways that professional geologists and lay people cultivated knowledge. Although he at times proved willing to accept facts that lay people contributed, he considered their reasoning abilities inferior. Unceasing demands for utilitarian knowledge that laymen could understand and translate into economic gain proved burdensome, undermined the survey, and eventually precipitated its demise. Business proprietors, company directors, and mining superintendents all possessed a "jealousy of professional and 'theoretical' interference with traditional and 'practical' usages."[37] This sentiment among Pennsylvania taxpayers became "a wave of suspicion and dislike" by 1842, eventually bringing the survey "to a dead stop."[38]

The inability of Rogers and the geologists who worked for him to accept geological science as a collaborative process in which they mediated local ways of knowing the environment with their own geological theories contributed to the survey's demise. State geological surveys were highly personal institutions that reflected the character of their directors. Rogers proved no better than his assistant Lesley at fostering respect among Pennsylvania's population and indeed developed a reputation for having an arrogant and dour personality.[39] Rogers's tactless approach destroyed any effort to make his research accessible to a public beyond the geological community. He and his brother William deeply hated politics and despised politicians who pandered to local interests, a trait that would have served them well in appealing to the business enterprises that could lobby for or against funding for their surveys.[40] Rogers offended local mining enterprises with some of his findings, prompting local geologists, miners, and investors to question the necessity of the survey.[41] The ambitions of geologists to pursue their personal scientific agendas ultimately proved too alluring for geologists who worked on surveys to take seriously contributions of the people who funded them and whose experiences in nature had been the impetus for the survey's creation.

TOPOGRAPHICAL SCIENCE

The science of topography and the related practice of map making differentiated geologists' ways of knowing the landscape from the local knowledge nongeologists produced. Through a combination of symbols, colors, shading, and other graphical representations, a topographical map represents the surface of the land in three-dimensional form. Such maps provide the advantage of conveying horizontal and vertical positions of terrain in relation to one another. Lesley distinguished himself as one of the nineteenth century's best topographical map makers. His pioneering innovations and use of this practice allowed him to capture and record the Pennsylvania landscape's unique attributes and form. He captured the state's "dynamic forms," especially the Appalachian Mountains, which stood before him "so grand" that they "excited a perpetual enthusiasm" and presented opportunities for "infinite research."[42] Pennsylvania exhibited a world of "natural forces," and through the use of topographical map making, Lesley proudly proclaimed that "we took possession of it."[43] The environment participated dialectically in knowledge formation with all humans who traversed it. Geologists practiced qualitatively unique forms of knowing the landscape by studying its topography and reproducing its contours in visual form. By taking possession of the land, geologists consciously placed their constructions of the landscape atop a qualitative hierarchy and placed nongeologists' conceptions at the bottom. Practicing science in this manner made Lesley and his colleagues "not mineralogists, not miners, not learned in fossils, not geologists in the full sense of the term, but topographers."[44]

Topographers felt their practice of depicting the environment made them clear and uncontested winners in contests with laymen for authority over who knew nature best. Although accomplished as a topographer in his own right, Lesley bestowed credit for introducing topographical science in Pennsylvania to another assistant on the survey, James D. Whelpley. Lesley heaped great praise upon Whelpley, contending that he was "the first perfect topographical geologist our science had."[45] Lesley overstated the case but qualified his point somewhat by explaining that lay people also participated in this practice.[46] Whelpley received assistance from "strong-minded miners" whose experiences surveying for coal contributed "stores of local knowledge," but in the end he had to "rely chiefly on his own genius . . .

for organizing the confused mass of unstudied phenomena before him into a consistent whole."[47] Lesley clearly recognized that multiple constituencies participated in the process of making the knowledge that explained nature's subsurface geology. In the case of Pennsylvania's anthracite coal mining district, "there were hot contentions of boss miners, proprietors, and speculators, over the underground complications."[48] Still, Lesley left no doubt who triumphed in these hot debates. These were "questions not to be settled" by the people who profited from resources "but solely by the geometrical study of the features of the surface. Topography was master of the situation."[49] Topographical map makers who depicted surface landscapes greatly empowered themselves by gaining entrée into the knowledge governing the earth's subsurface geology. They demonstrated that topographical and geological sciences complemented each other in expressing relationships between the earth's surface and subsurface.

Lesley's conception of topography as both a science and an art underscored how the maps he produced possessed both utilitarian and aesthetic attributes. On one hand, topography was "like every other science" because it conveyed how a few basic laws of nature "express themselves upon the surface of the earth."[50] However, the best topographers approached their craft as rational scientists and as artists who pursued "truth" in the landscapes they sketched. To pursue truth in map making meant to include as many details as possible. Lesley could not provide an exact formula to specify how to balance science and art in a map because laws governing relationships between the two were "not conventional nor empirical."[51] He could only advise that the artist "put himself in true relations with this grand mute object of his study, and learn its own record of its wonderful experience."[52] Learning to the read the surface of the earth was the first lesson geologists confronted. Making maps of the surface was "indispensable to the progress of discovery" and represented "the crowning achievement of a perfect knowledge of its geology."[53]

Lesley aspired to provide a comprehensible methodology for creating consistently reliable maps, but his discourse on topography revealed how geology's visual language potentially represented the geologist's perspective more than nature itself. Some map makers interpreted nature incorrectly, and Lesley warned of "mischievous" landscape sketches that "distort" nature's form and "*oblige* deductions to be wrong."[54] Maps made carelessly while riding horseback or standing on a railroad embankment offer "mere guesswork" to fill up space. To ensure quality maps, policing agencies such as a normal school or a board of control were "absolute necessities of the profession."[55] Lesley believed that his growing expertise in making topographical maps provided him with the credentials to police the quality of maps on the first survey, and he would scrutinize maps even more closely when he was appointed director of Pennsylvania's second geological survey. Still, Lesley acknowledged that capturing the earth's geological complexity was a very subjective endeavor, and "no one, in fact, seems to know what a perfect map should be." Nevertheless, maps and other visual media had become indispensable to the practice of geology, prompting Lesley to ask, "What is a geologist — *what is his geology* without one?"[56] Maps and the ability to make them distinguished geologists from other practitioners who generated knowledge about nature. Only geologists could make maps, Lesley contended, because "the geologist alone looks for *surface indications of internal structure*."[57] Topographical maps potentially possessed a great deal of power when they depicted geologists' theories explaining relationships between surface and subsurface geology. Of course, not all geologists theorized similarly, and individual maps could function as highly idiosyncratic cultural artifacts.

Maps, images, and other elements of the visual language central to the emerging field of geological science sometimes elicited fierce battles for professional authority among scientists. The issue of who received credit for creating a geological map lay at the heart of a dispute between Lesley and Rogers that erupted in 1858 when Rogers published the second and last volume of his long-delayed final report of the Pennsylvania survey. Although he identified the assistants who had performed field work by name, he did not specify the degree or type of each assistant's contribution. Lesley considered the list of names insufficient attribution and accused Rogers of "the most extensive scientific theft of the present age": "in this immense work of nearly two thousand pages, magnificently illustrated with maps, sections, and pictures of all kinds to the number of nearly a thousand, are the results of the toil of many men for many years, all appropriated by one man to himself."[58] Lesley resented Rogers's practice of collecting assistants' field notes, replicating them in his report, and conveying the impression that "he has done it all." Notably Lesley focused his ire upon the "magnificently illustrated" maps because he spent much of his time at the survey producing them, excelled at the activity, and felt that the images best showcased his professional abilities. He grew infuriated that Rogers "gives credit to no one else" and suggested to "the world" the maps were his alone. Rogers had "erased from the map of Pennsylvania" Lesley's name even though he "alone constructed and compiled" it.[59] As young scientists working to establish themselves professionally, assistants felt the mere mention of their names failed to acknowledge sufficiently their accomplishments.[60]

William Rogers defended his brother against Lesley's accusations by arguing primarily that the nature of scientific practice required scientists to assist one another, if not actively collaborate. William's defense of Henry included three central points: multiple geologists performed work resulting in the map; Rogers ultimately possessed sole authority over this knowledge, regardless of who generated it; and Lesley as well as the other assistants had in fact received credit for their efforts. William explained that Henry had mentioned "by name and title" each assistant who worked for the survey from 1836 to 1841.[61] Henry held a "high appreciation" for their "zeal and ability" and admired their "energy and fortitude in confronting difficulties" involved with their work.[62] To detail each of the "special parts performed by the individual members" of the survey, however, would lead to a "tedious" narrative simply because "their duties were so multifarious."[63] Far from failing to give credit, the Rogers brothers considered their assistants necessary to completing their work and relied on them greatly.[64] William rebutted each accusation but especially the claim that Lesley had solely authored the geological map in the final report. William stated emphatically that "almost every assistant" Henry hired to work on the survey "contributed, in one form or other, and in a greater or lesser degree, to the data embodied on this published map," making Lesley's claim of sole authorship "rather extravagant."[65] The dispute revealed how central the visual language of images had become to geological science and the accelerating stakes involved in claiming authority for their creation.

Lesley issued other protests at what he perceived as additional attempts by Henry Rogers to appropriate geological images. In addition to the geological map, Rogers also claimed credit for several cross-sections Lesley had produced. A cross-section is an image that depicts from a side perspective how strata deposited within a portion of the earth, as if viewing a slice of chocolate layer cake. Lesley argued that he "was the first to propose and alone executed" thirteen geological cross-sections and "all the vertical sections of the coal measures" for which Rogers gave him no credit.[66] As was the case with the geological map, William noted that "every member of the survey" created vertical cross-sections based upon their field work."[67] Thus multiple geological assistants contributed information Rogers relied upon to create geological images.

The visual images that resulted from and contributed to professional geology's emergence as a distinct discipline originated from a variety of cognitive sources. Sections represent "thought-experiments" and embody extrapolations derived from theoretical speculations generated during field work.[68] A geologist who conceived of theories that accurately explained configurations of subsurface geology possessed a great deal of authority if he could depict his theories in visual form as cross-sections. Lesley's claim that he originated cross-sections as a visual form elicited the Rogers brothers' disbelief. "We are at a loss to understand," William lamented, "how Mr. Lesley can claim to have proposed a form of illustration that is almost as old as the science of geology."[69] Lesley's claim of sole authorship and of originating the use of cross-sections reflected his claim for the intellectual power those images represented. Over time, geological images bore an increasing load of theoretical meaning.[70] The more theoretical power they possessed, the greater the incentive to declare possession of them.

Lesley's statements confirm that even though he may have solely authored specific geological images, he derived their final form by relying upon other geologists' intellectual work. He simultaneously claimed sole authorship of hundreds of images while admitting that he "re-drew" them based upon his field notes or "from the fine diagrams of Whelpley," who had worked as the survey's chief topographer.[71] Other images Lesley "re-drew" but for which he claimed sole authorship were based on "the less perfect pen and pencil sketches of other members of the survey."[72] Like Rogers, Lesley borrowed from others to produce visual images but did not consider his actions "scientific theft." Though he conceded that Rogers's final report contained "a world of necessary facts," he selectively recognized collaboration as a means to produce geological knowledge. Lesley craved professional authority as a geological assistant and refused to recognize that his employment with the survey was based on collaboration.

SECOND PENNSYLVANIA GEOLOGICAL SURVEY, 1873–89

Ironically Lesley instituted a model of collaborative research when the discovery of oil in Pennsylvania prompted the legislature to create a second geological survey and appointed him its director. The discipline of geology had changed greatly from 1838 to 1873, years that mark creation of the first and second Pennsylvania geological surveys. As a young man Lesley lacked sympathy with Rogers's reliance on collaboration in order to fulfill the first survey's administrative demands. However, Lesley had matured, gained experience, and advanced his professional standing in the years since he had worked for Rogers. Faced with similar administrative and political pressures to produce information that taxpayers and legislators deemed practical, Lesley confronted the reality that only by surrounding himself with geological assistants who could conduct the extensive field work could he fulfill the survey's mandate.

Lesley's willingness to collaborate extended to assistants but not to laymen, whom he continued to regard as superstitious, poorly educated, and ill suited to think geologically. The disregard he had exhibited for laymen's ideas about how and where to find natural resources accelerated when he assumed the directorship of the second geological survey. The first report he issued was a history of geological exploration in the state that he hoped would educate the board of commissioners about how surveys functioned and "to show what a Geological State Survey really means."[73] Lesley declared that a survey possesses the "power to stimulate the intellect of the State" and that geologists could "sweep away costly superstitions respecting the mineral resources of the Commonwealth."[74] He saw geologists as heroes who should enlighten the masses. Geologists had been actively engaged in Pennsylvania for forty years, he explained, and citizens of the state "ought to have exhibited to them the manly struggle" these scientists had endured. Their struggle was a "contest" between a "youthful, growing, strengthening and maturing science" and the "prejudices and falsifications of half-educated men."[75] Lesley's historical account reads like a morality play. Geologists functioned as "manly" heroes while nonexperts were unreasoning, "half-educated" obstacles who exhibited superstition, prejudice, and false ideas.

Geological surveys required staffs of experts trained in geology and engineering, and these qualifications ruled out laymen regardless of how much oil they had found. Lesley explained to residents of a small town outside of Pittsburgh the merits of the assistants who worked for him when townspeople demanded that the survey hire more people living in the town. Writing to the local newspaper, Lesley informed

Monongahela's citizens that "geology is a special science and requires specially trained experts." He frequently encountered people who "think themselves and are judged by their friends and neighbors" qualified to work as geologists, but their reports "would be good for nothing" and possess "little or no value." Lesley's experience working for Rogers on the first survey and later as a private consultant taught him that "good geological work can only be done by men long trained to it." Anyone who had "not been engaged for years in practical geological field work" understood "how few such experts" qualified for employment on the survey. To meet Lesley's standards, applicants needed training for at least a year before he deemed their geological work reliable. Regardless of how much experience a prospector possessed, "however enthusiastic his so-called love for geology," a prospector without training qualified as merely a "raw hand" whom Lesley refused to take seriously.[76]

Residents in and around Pittsburgh resented Lesley's dismissal of their knowledge of resources; they wanted influence over the survey. The editor of the Monongahela newspaper expressed irritation over the abundance of assistants from Philadelphia working for the survey and the lack of locals who resided in Pennsylvania's ten western coal-producing counties. Lesley "pretends" that he could not find qualified locals to hire, the editor argued, but "we consider this an unjust reflection upon the ability and spirit of our engineers." The city's residents felt professionals from Pennsylvania's metropolitan center to the east disparaged westerners' experience and intellects. The editor asked, "Does Pittsburgh so lack brains? Has Philadelphia such a super-abundance?" He contended that Lesley "dwells too much on the presumed ignorance of the people" and that "other folks" had their own ideas of how to produce information.[77] Lesley interpreted this complaint and other disapproving editorials published in Pittsburgh as "reckless" evidence that "only go to show how completely the average business man misunderstands and necessarily misunderstands the real objects, drift, and necessary methods of a state geological survey."[78] Pennsylvania residents whose livelihoods depended on natural resources felt a strong attachment to the environments in which they lived and worked and had successfully produced reliable knowledge. Meanwhile, geologists found the information these residents generated useful but only as factual building blocks to construct more powerful theories with the capacity to explain how resources accumulated translocally and benefited wider constituencies.

Geologists generated the best information when they established relationships between theories and the data they gathered conducting fieldwork; Lesley felt that laymen lacked the capacity to correlate facts with theories. He did not object to theories per se, as long as "they are supported by a great multitude of harmonized facts."[79] As he saw the matter, "so-called" practical men exhibited no practicality at all in formulating theories about how to find oil. Instead, it was the geologists who produced more reliable information because they "base their theories on a wide range of well connected facts" whereas practical men "establish theoretical prejudices" based upon a "narrow circle of the facts."[80] Simply put, geologists established more "harmony" between fact and theory. Lesley did refer to practical men's use of the "belt-line theory" to find oil, although he did not identify it by name. Practical men succeeded when they applied the theory because doing so sometimes yielded oil but only in particular locales, and they failed when applying the same theory to areas with different geology. Lesley did not want geologists working for him to make similar mistakes, and he took great pains to ensure that the assistants he hired theorized with caution.

Lesley so fully embraced collaboration with properly trained assistants that he developed pride in mentoring young men eager to establish their scientific careers. He encouraged his nephew Benjamin Lyman, who worked as an assistant for the survey, to exercise tolerance and understanding in his professional relationships. Lesley explained that he had supervised many men, "young and old," and had learned "to exercise a great deal of patience."[81] Colleagues must "make the best of each other," and he urged Lyman to consider that "there is more good in every young fellow than appears at first."[82] Learning this lesson did not come easily for Lesley. He conceded that "many a harsh word I have said which I was sorry for," and he had at times behaved in an "unjust and despotic" manner.[83] Altruism motivated his despotism, he claimed, for the "chief desire was to advance their interests and consult their comfort rather than my own."[84] As director of the survey, Lesley had learned a lesson that would have served him well when he worked as an assistant on the first geological survey. Collaboration required hard work and a significant degree of patience when working with young men still learning the rudiments of geological science.

Lesley's patience for collaboration faltered when assistants did not correlate facts and theories as judiciously as he would have liked. He demanded that assistants' reports contain only "simple descriptions of work done, records of facts observed, and explanations of the local geology" and describe only "what is *known* by geologists."[85] Simply put, the assistants must refrain from "discussion of abstruse questions" because they were "still subjects of speculation among geologists."[86] Giving his assistants too much intellectual freedom would have undermined his mandate to generate information that helped people locate resources.

The pressure to generate practical information in a timely fashion prompted Lesley to scrutinize assistants' theoretical statements as closely as possible. As editor of all the survey's published documents, Lesley proofread reports his assistants wrote based upon the field work they performed. He committed himself to "publishing results as fast as obtained" so as to appease the board of commissioners who appointed him.[87] Even with the demands of speedy publication, he closely monitored assistants' theoretical statements. At times he eliminated conclusions he considered too theoretical because they lacked sufficient factual proof.[88] By his own account he expended "unceasing labor as an editor, every day of the entire year. Every sentence of every report must be revised."[89] Assistants remembered his editing with a mixture of admiration and dread. One assistant recalled that "in several cases he rewrote reports" in order to present findings "in a more systematic way."[90] When Lesley detected an error or reasoning that overreached, he "relieved himself in a communication which was a model of terseness and clearness" and unleashed criticisms that "were none too mild."[91] Initially, Lesley's editorial notes "did not always seem to the authors to be either necessary or valuable. Yet, after a score of years, it must be conceded that not a few of the suggestions, which were most unsatisfactory at the time, have proved to be of lasting value."[92] His nephew Lyman also remembered that his uncle performed "a great deal of conscientious editorial work" and entirely rewrote reports.[93] Images such as maps and illustrations also "passed under his close scrutiny and revision."[94] However obsessive and draconian Lesley's approach may have appeared to assistants, he understood that young geologists needed to learn that knowledge consisted of both facts and theories judiciously balanced in relation to one another, and he indoctrinated most of them with this viewpoint.

One of his assistants, Henry M. Chance, watched his boss at work and witnessed his ability to maintain the precarious balance between fact and theory. Chance observed that Lesley often displayed a great fondness for theoretical science. According to Chance, Lesley was "intensely interested in abstract science, loving it for itself alone."[95] Lesley possessed an imagination well suited for conceptualizing how geological processes unfolded over wide expanses of time. Chance remembered Lesley "dreamily looking back through the ages, reconstructing mentally the conditions and forces at work, which have given us the earth as we now have it, and perhaps looking forward to foretell the future."[96] Despite such reveries, Chance also considered Lesley "eminently practical, a man of affairs, an engineer." Lesley's mind often shifted abruptly from theoretical to practical matters. On one hand, he contemplated geological processes unfolding over millennia, but "in a moment, divorcing these poetic dreams, he became a utilitarian, a conservative mining engineer, accepting and weighing only those facts and agencies having direct bearing upon the extent, quality and value of the minerals with which as a master of the art he continually had to deal."[97] Lesley conceived of science as an exercise requiring both applied and theoretical insights and expected the same from his assistants.

Lesley quickly silenced one of his assistants, Israel C. White, who failed to challenge his own assumptions and indulged too freely in theoretical speculations. In his reports White frequently violated Lesley's sense of appropriate geological reasoning. He worked for Lesley from 1875 to 1883, and their professional relationship did not always prove easy.[98] Lesley particularly disapproved of a report White had written on the geology of Mercer County. "His style is so verbose and invested that I will not accept it," Lesley complained in his diary.[99] He criticized White for making unwarranted speculations, "assertions and generalizations which I will not allow."[100] Lesley rewrote 124 pages of White's manuscript but revising the entire 250 pages proved "too heavy a job," and he resolved to "delete doubtful assertions" and publish the remainder in its original form.[101] He had rewritten other assistants' reports, but White's rampant proclivity for theorizing violated Lesley's most cherished maxim.

Even when he praised White's theoretical insights for their systematic rigor and universal application, Lesley reminded him that local knowledge obtained through field work made such insights possible. In one instance White had "done nobly" to identify a relationship between "the Ohio rocks and our oil belt," but this good geological work resulted from "sticking to your county work, and doing it minutely and locally."[102] Only a seasoned veteran could understand how thoroughly speculations unsupported by careful field work translated into bad science. Lesley reminded White that "prediction and speculation on insuffic[ient] data have been and still are the curses of our science, and when you are 60 years old as I am next month you will feel this keenly."[103] Unsubstantiated speculations, especially in the oil industry, provided opportunities for unscrupulous speculators to take advantage of investors. In this climate scientists had to fight to win professional authority and credibility. The price they paid for this victory was the field work they performed that furnished the data to substantiate their theories.

Tensions particularly arose when Lesley suspected that White formulated theories to enhance his professional authority within the scientific community rather than to aid the citizens of Pennsylvania. A delay in submitting a geological report aroused Lesley's suspicion and prompted him to state unequivocally: "The treatment of *general* geological subjects is not called for. I care nothing at all about the 'geologists of the country.'"[104] Professional geologists had waited forty years since the first survey ended for additional geological information and could continue to wait as far as Lesley was concerned. He reminded White that they wrote reports "not at all for the geologists of the country but for the citizens of the counties and the state who pay for the Survey."[105] The same issue resurfaced the following year, prompting Lesley to declare that "in one respect I shall be despotic."[106] Where there existed even a shadow of a doubt, "I will not permit a *confident expression* of *a general character* to go into type."[107] The issue of scientific authority was at stake in the matter of publishing geological theories in the reports. Lesley understood that if theoretical statements issued under the cloak of the survey later proved to be false, not only White's authority would suffer but so too would the survey's and his own as its director. He refused to have White's report "quoted as absolute authority" on any issue for which evidence was not "ample and indisputable."[108] Lesley thus protected his professional reputation and the survey's against young geological assistants who overreached in their reports. When evidence was lacking, he told White, "I shall change your *general assertion* into a *personal assertion* and let the statement go on *your authority alone*."[109]

Henry Chance was an assistant who appears to have learned his mentor's lessons well and unleashed similar invective on practical men for applying geological theories too dogmatically. Chance conducted field investigations in Clarion County, one of Pennsylvania's richest in oil resources. Like other geologists who worked for Lesley, he discussed the merits of the belt-line theory in a report he wrote for the geological survey. He did not dismiss the theory outright, conceding that prospectors could rely in "great measure on the *general trend* of the productive areas" to locate future well sites.[110] The belt-line theory served a purpose "*within certain limits*" and when applied "in a general way, the idea is a good one."[111] However, the notion that streaks of oil "run in unvarying straight lines" and can be traced for several miles greatly oversimplified the geological complexity of oil accumulation. Frequent successes in its application had converted "hundreds of producers to an unreasoning, dogmatic belief in this theory."[112] Chance lamented practical men's lack of geological sophistication. He argued that the theory's limitations were "self evident to anyone familiar with the character of sedimentary rocks and the agencies by which they were deposited."[113] Nature possessed agency and the best prospectors strove to understand how geological processes reflected that action. Practical men who did not consider nature's contingency might find oil in some locales, but the odds worsened the longer they refused to extrapolate larger geological principles from local knowledge. Chance appealed to the biases and instincts of Lesley in pointing out the naiveté of practical men who thought they could impose their will onto nature by dogmatically applying the belt-line theory to discover oil. The notion of using a compass line "to trace out one individual sandy streak is, as Prof. Lesley has described it, like a boy marking pencil lines on the top board of a wood pile to determine the direction of the grain in the bottom board."[114] Just as grains in the boards of a wood pile were unrelated, surface geology did not necessarily illuminate subsurface geology. Geologists who theorized too freely about buried strata, or boards not visible from the top of the wood pile, potentially espoused inaccurate information and risked their professional reputation.

Henry E. Wrigley, another Lesley assistant, also noted practical men's poor theorizing and, although conceding that their prospecting methods succeeded at times, argued that ultimately these methods proved inadequate because they failed to account for nature's agency. Belt-line theory oversimplified nature and reflected an attempt to superimpose humans' ideas onto intricate geological forces that required additional ongoing study. Wrigley expressed this view in a report he prepared on the Pennsylvania oil industry. Commissioned by Lesley, Wrigley had worked as an architect and civil engineer rather than as a geologist. His report revealed the perspective of a professional engineer trained to consider how his designs should accommodate physical boundaries imposed by the natural environment.[115] In Wrigley's view practical oil men applied their supposed "lines" with no systematic rigor. They envisaged straight lines drawn from one productive well to the next to serve as a guide for finding oil, but in actuality their lines meandered across the landscape. Wrigley observed that if he conducted an aerial study of belt-lines from a hot-air balloon, it "would probably show them to be as accurate as the course of a drunken elephant through the jungle."[116] At another point he characterized this method of prospecting as crude.[117] Meandering lines suggested that practical men gave very little consideration to the relationship between their prospecting methods and subsurface geology. Wrigley considered Cyrus Angell's application of the belt-line theory "the most valuable of all these propositions," but it too failed to account for the complexity of subsurface geology. Upon perusing a map, Angell noted that productive oil wells existed along "a straight line" that ran sixteen degrees in a northeasterly direction. In Wrigley's view Angell, however, ultimately "failed to establish" a useful theory for prospecting because his conception of how oil accumulated contained "the grain of truth to the pound of error." Angell failed to see "the error" of his ways by failing to consider "the fact that nature never works with absolutely straight lines."[118] An engineer like Wrigley and the geologists for whom he worked understood that prospecting in a straight line might lead to some oil strikes but that the contingency and complexity of the earth's geology forced those seeking oil to constantly challenge their assumptions.

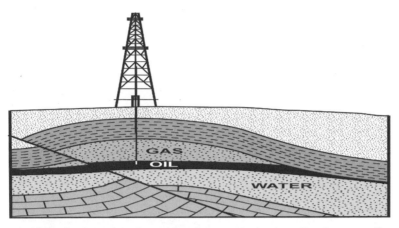

3. An idealized cross-section of an anticline demonstrating how gas, oil, and water stratify according to their densities. A topographical map depicting an aerial view of the anticline offered a useful strategy for knowing where to begin drilling. Courtesy of Pennsylvania Geological Survey.

THE ANTICLINAL THEORY

Two years after leaving the survey, Israel White demonstrated to prospectors the practical application of a theory for increasing the probability of finding oil in some locales. Oil speculators expended large sums of money drilling holes and wanted their investments to yield oil. A theory that could significantly increase the probability of finding oil before drilling began offered practical and economic advantages that presented the oil industry with a reliable and systematic prospecting method. Although he was not the first to articulate the anticlinal theory, White published an article in which he explained how he had successfully located oil through its application in West Virginia.[119] Described in its simplest form, an anticline is a geological structure that consists of upfolded strata and may resemble an elevation in the earth's surface such as a hill. Some of these structures, if not buried by other layers of earth, can be observed at the surface of the earth. Oil accumulates within them as it migrates upward through permeable strata and becomes trapped beneath an impermeable shale stratum making up the anticline's roof. Gas and water also migrate

4. One side of the Coalinga anticline in the Eastside field, Fresno County, California, 1907. Note that erosion of soft sedimentary beds created steep slopes, presenting difficulties for producing oil on this geological structure. Courtesy of United States Geological Survey.

into the anticline, but given the differing densities, oil, gas, and water segregate into three distinct layers from top to bottom: gas on top, oil in the middle, and water on the bottom. By surveying the anticline and measuring the midpoint of its elevation, White showed how to eliminate some of the guesswork involved in locating oil housed in anticlines.[120] Applied strategically, the anticlinal theory proved effective and efficient because drilling into the structure at the precise place where oil accumulated maximized production when pressurized gas and water pushed it to the surface. Other geological phenomena also trapped oil, but anticlines account for eighty percent of the world's discovered oil and gas.[121] White's successful application in 1885 of the theory had the potential to revolutionize the oil industry.

No revolution occurred, however, because most of Pennsylvania's oil accumulated for geological reasons unrelated to anticlinal structures. The anticlinal theory eventually affected the oil industry in other states and at a later date. The theory proved less useful in Pennsyl-

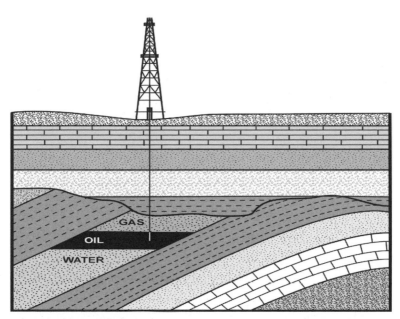

5. Stratigraphic trap. Oil accumulates in anticlines because of the impervious shale layer atop the geological structure, but oil in stratigraphic traps accumulates because of a discontinuity in the stratum in which the oil is traveling. Courtesy of Pennsylvania Geological Survey.

vania, however, because most of that state's oil accumulated due to a geological phenomenon called a stratigraphic trap. Unlike structural traps such as anticlines, oil accumulates in a stratigraphic trap because of a discontinuity or variation in the strata.[122] These discontinuous strata, or unconformities, were not observable at the earth's surface and therefore required highly innovative geological theorizing and subsurface geological mapping to understand their relationship to the oil they held.

JOHN FRANKLIN CARLL

In nineteenth-century Pennsylvania, by systematically gathering data to build subsurface contour maps that illustrated geological theories and engineering principles that the industry built upon throughout the twentieth century, John Franklin Carll first articulated how stratig-

6. John Franklin Carll. Courtesy of Pennsylvania Geological Survey.

raphy trapped oil. Many geologists interested in the history of the oil industry hail Carll as the father of petroleum geology and engineering.[123] Lesley hired Carll to oversee most of the work involving oil on the geological survey, but the latter did not begin his professional life as a geologist. Born in Bushwick, New York, in 1828, Carll studied civil engineering and worked as an engineer before starting his own wire manufacturing business. After experiencing the death of his wife and two children and later a fire that destroyed his business, Carll at thirty-six departed New York and headed for the Pennsylvania oil fields. Drawing upon his engineering experience, he professionally reinvented himself as an oil producer by carefully observing different rock strata at drill sites and collecting well records. His professional expertise consisted of a mélange of theoretical and factual traditions. Carll cultivated the practice of close hands-on observation in order to learn as much as possible about geological formations; he developed a widespread reputation as a geologist so that many oil producers solicited his advice and expertise about drilling at prospective sites.[124] Carll's approach to knowledge production illustrated perfectly how the best oil prospectors followed no single epistemic plan and carefully crafted relationships between their experiences conducting field work and the theories they fashioned to explain oil accumulation.

Many of Pennsylvania's earliest drillers frustrated Carll's effort to gather as much data as possible because they kept records carelessly and unsystematically if they bothered to keep them at all. This neglect complicated Carll's efforts, although he recognized that the survey could not fulfill its mission without the local knowledge drillers possessed. He wrote that drillers kept records so poorly that "to the ordinary reader" their data indicating thickness of sand, depths of wells, surface elevations and undulations at the surface were "simply unintelligible."[125] He complained, "How to secure well records in a complete and reliable form has been one of the perplexing questions of the Survey."[126] The rush of excitement that accompanied the discovery of oil militated against drillers and owners recording geological data. Science suffered because oil interests focused on short-term gains that reaped "the largest margin of profit."[127] The "whirl of excitement" that pervaded a new and productive oil field distracted "even the most staid and methodical student of nature," leaving "no time for scientific inquiry."[128] Carll recognized that drillers possessed valuable firsthand information about geology they observed at well sites but that their voluminous records and samples needed to be organized more systematically to transform their data into geological theories that explained more clearly how and where oil accumulated.

Even though Carll valued information he gleaned from drillers' logs, he too felt that practical men constructed theories that correlated poorly with the local knowledge they uncovered through drilling. In attempting to "get a correct idea" of how oil-bearing sands stratified geologically, he relied upon well records "given promiscuously" by various drillers and well owners. The data was better than nothing but remained "colored by their individual theories or pecuniary interests."[129] Drillers and geologists generated different kinds of knowledge because their encounters in nature provided them with distinctive blends of intellectual and sensory experiences that reflected their unique epistemic goals. Carll did not dismiss laymen's knowledge quite as harshly as Lesley, but he recognized the limitations of their geological judgments. For a geologist to depend on a driller's judgment was "a delusion and a snare."[130]

Carll depicted theories of oil accumulation in maps but to do so required collaboration with drillers and well owners throughout

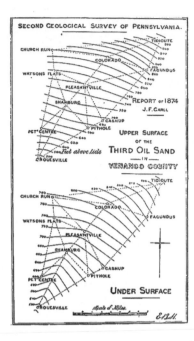

7. Carll's subsurface structure contour map. This image depicts the underground contours of the oil-rich third sand in Venango County, Pennsylvania. Note that the depth of the strata decreases from 780 feet to 520 feet from Church Run in the north to Rouesville in the south. Knowing the progressive dip of the strata facilitated drillers in finding oil. Carll, *Report of Progress*, 17–18.

Pennsylvania in order to gather as many samples as possible. Through these efforts he produced by the 1880s some of the most significant petroleum-related geology and engineering knowledge to that time. Carll began organizing efforts to gather well samples by assembling a producers' association and hiring assistants to monitor fields and gather samples. Carll's identification and mapping of the oil-bearing strata known as the Venango Third Sand was one of his most important achievements resulting from these efforts. Drawing upon well samples and his own field work, he constructed and published the first contour maps of subsurface oil-bearing formations in North America.[131] These maps illustrated that oil accumulated in subsurface stratigraphic layers of sand and not for reasons related to structures such as anticlines, which prospectors could often observe at the surface without the aid of a map.[132] Whereas previous theorists like Israel White advocated the search for structures, Carll demonstrated that in Pennsylvania structures were less important than stratigraphy. Carll's maps and the accompanying report represented a significant practical and theoretical achievement. From a practical standpoint, he provided drillers with a more reliable guide for navigating Pennsylvania's geology rather than merely following the belt-line theory. Theoretically Carll demonstrated that oil — especially in Pennsylvania — accumulated for reasons related to geological stratigraphy rather than to geological structures.

Collaboration to formulate knowledge for finding natural resources never came easily for the geologists and laymen of Pennsylvania. Although most constituencies in the state wanted to locate resources, they possessed different epistemic missions in their searches that at times erupted into disagreements and conflicts that stymied their efforts but also generated useful knowledge for locating oil. As a young scientist bent on establishing his professional credentials, Lesley contributed to these disputes, but in doing so, he demonstrated the enormous premium arising from an individual's encounters with nature. He felt the geological maps he produced represented more than just contributions to science but functioned also as personal testaments to his experiences and understandings of nature. Thus he guarded these maps jealously. Topographical maps offered snapshots into Lesley's mind at work, conceptualizing geological forces as they impressed themselves onto his senses.

As personal and individualistic as maps often were, however, geologists relied upon lay people and their peers within the professional community of scientists for information to generate them. Lay people provided firsthand observations, estimates, and ideas about quantity and quality of resources they encountered while working in nature. Fellow geologists likewise generated knowledge conducting field work that survey directors compiled to draft their final reports and maps. In short, all these parties collaborated to conduct science. Collaboration was the central reason, of course, for creating surveys in which a selected group of practitioners could efficiently traverse the state and record their observations in as comprehensible a form as possible. Carll collaborated better than anyone in his efforts to construct geological theories that explained how oil accumulated in Pennsylvania. He was the exemplar of collaborative science because he fashioned sound geological theories by taking seriously the knowledge laymen generated, and he created maps to depict a subsurface environment no person had ever witnessed.

2 Contested Knowledge

3

Shared Authority

PRACTICAL OIL MEN AND
PROFESSIONAL GEOLOGISTS

Roswell Johnson set up shop as a petroleum geologist in the northeast corner of Oklahoma in 1908 and began advertising his services as an independent consultant. After arriving in the small town of Bartlesville, he began dutifully running ads and occasionally even publishing articles in the pages of the petroleum industry bible, the *Oil and Gas Journal*, as part of a campaign to convince oil men of the practical advantages his services could provide.[1] Prospectors had been pouring into the midcontinent region from the Appalachian oil fields, and Johnson perceived an untapped market for his consulting skills if only he could demonstrate that geology offered a better prospecting method than vernacular beliefs such as the beltline theory. He explained that even though practical men had successfully found oil in Pennsylvania by drilling along lines that ran at forty-five-degree angles, "many operators had accordingly come to believe that there was some mystic power in this particular direction" and that they could find oil in Oklahoma by applying the theory.[2] When this approach failed to produce oil in the midcontinent fields, "those mystically inclined" adopted a new faith that drilling along a line of 22½ degrees would yield oil.[3] Johnson contended that geology, and particularly the anticlinal theory, offered a better prospecting tool than the belt-line theory. Although his advertisement sat prominently on the corner of the same page on which one of his article appeared, he had attracted so few clients by

1912 that he accepted a position teaching geology at the University of Pittsburgh and quit the oil business.

Sensing that he championed a losing cause, Johnson left Oklahoma, but his timing could not have been worse because within a year after his departure, the oil industry began seeking geologists' advice on an unprecedented scale. Some practical oil men accepted petroleum geology earlier than others, but 1913 proved a pivotal year in the history of the oil industry because many who had resisted geology began to take it more seriously.[4] During this year geologists mapped an anticline in Cushing, Oklahoma, and demonstrated with a visual representation that they could find oil by applying geological principles. Because geologists could never say with absolute certainty whether drilling in a particular place would strike oil, their techniques and ideas remained open to debate. Practical men still found their share of oil, but they began to offer more specific information about the physical conditions beneath the surface of the earth that they used to create visual images of the strata, thereby eliminating much of the guesswork.

Johnson's experiences indicated that the contests for the authority and power to interpret the meaning of local knowledge remained divided as the oil industry shifted westward from Pennsylvania to the southern plains states of Kansas, Oklahoma, and Texas. The Pennsylvania survey provided a unique forum, unprecedented in scale and scope, to formulate and codify geological knowledge in order to find oil, but it met with mixed results. Fractures among practitioners who understood their relationships to nature differently from one another complicated that effort as these practitioners carried their contested understandings of work and nature with them to the southern plains. The lines between the knowledge different practitioners cultivated continued to overlap, and as the southern plains environment yielded great quantities of oil, the contests for authority to determine how geological knowledge revealed those oil reserves intensified. The environment participated in those contests by requiring prospectors to cultivate new forms of local knowledge in order to understand how this region's geology differed from states farther east. The lines between natural and cultural landscapes blurred on the southern plains as they had in Pennsylvania, and practitioners' negotiations of those boundaries created tensions out of which the field of petroleum geology would emerge.

The oil industry moved west over several decades as the United States underwent dramatic changes. From approximately 1880 to 1920, the nation gradually transformed from mostly autonomous and localized island communities to a distended society of centralized planning with institutions that aimed to rationally organize science, industry, and government. John D. Rockefeller and his Standard Oil Company stepped into American business's organizational void to eliminate the chaotic production practices that resulted in too much oil and destabilized markets. Standard's imposition of order resulted in a market monopoly that limited economic opportunities for practical men and stifled the individualist ethic that animated their work habits and identities. As the flow from Pennsylvania's wells tapered off, pioneering oil men who felt their freedom diminishing fled eastern locales for opportunities in the West's emergent island communities.

Although Americans' search for order greatly affected all facets of their lives, the transformation occurred gradually and at times encountered resistance.[5] Venturing west offered practical oil men an opportunity to protest Standard Oil's octopus-like stranglehold over their lives. In the West oil men initially explored for oil with methods they had long cultivated through hands-on practice and trial and error. Lacking the capital to participate in an increasingly larger economy of scale, practical men often resented the presence of geologists who, employed by large corporations, began articulating an increasingly arcane knowledge of geological theories they postulated but could not always verify. Practical men who operated on a shoestring budget prided themselves on their practicality, which meant performing as many of the duties to produce oil as possible including prospecting. Geologists found oil, however, and their presence signaled that new organizational forms and larger economies of scale presented significant competitive advantages to companies who could afford to hire them. The 1911 antitrust ruling that broke apart Standard's monopoly complicated its efforts to access western markets. Standard proved willing to embrace science and technology to build an efficient organization, but it adopted geological prospecting later than oil companies in the West. Standard's strategy had always hinged on taking advantage of surplus production, and the company saw no need to expend capital to produce more oil when a glut already existed. In the West after 1911, new oil corporations emerged who increasingly adopted petroleum geology. Corporate leaders began taking notice of Israel C. White's 1885 demonstration of the anticlinal theory, and in succeeding decades they hired geologists for their professional expertise. Still, they would not begin hiring geologists en masse until approximately 1913, at which time geologists began asserting more control over the oil finding process.

As pivotal as the mapping of the Cushing anticline was to petroleum geology's acceptance, to suggest that the oil industry suddenly embraced a more scientific approach greatly overstates the case and understates and misrepresents practical men's contributions to the knowledge necessary for finding oil. Many practical men still resisted geology even after production began at Cushing because they continued to find oil with traditional methods such as the belt-line theory. Their knowledge represented a stochastic mode of reasoning better characterized as experienced intuition rather than haphazard guessing uninformed by scientific principles.[6] Many felt no need to adopt geology because they had long-crafted novel solutions that solved practical problems amid daunting uncertainties the search for oil presented.[7] Geologists strove to reduce the level of uncertainty by gleaning important data from practical men's drilling logs and using it to formulate geological theories about where to find oil. Although maps of the Cushing field proved pivotal in convincing many in the industry of geology's utility, no clear line demarcated the traditional prescientific knowledge that prospectors used to find oil from the geological principles the industry began to adopt in 1913 and that eventually revolutionized the search for oil.

TOM SLICK

Tom Slick offers an excellent example of a Pennsylvania practical oil man who followed the industry westward to Oklahoma and who, like his contemporaries, often found oil by relying upon his senses. After finishing high school, Slick began working for his father, who owned drilling rigs in various Pennsylvania fields.[8] Slick suggested that his birth in the heart of oil country and to a father who worked as a driller instilled within him an innate sensory relationship to oil: "I came west from Clarion, Pa., where I was born among the oil derricks. The first sniff of air I ever breathed into my nostrils was laden with the odor of oil."[9] One early historian of the oil industry seemed to accept Slick's suggestion that he possessed an extrasensory power and could literally smell oil as deep as two thousand feet below the surface.[10] Like a bee drawn to a flower, "the smell of oil sands was perfume to his nostrils," and this allegedly innate gift supposedly explained Slick's success.[11] By 1903 production in Pennsylvania and other fields east of the Mississippi River began to decline, and Slick, his brother, and their father migrated to Chanute, Kansas. Like others who headed west in search of oil, the Slick family carried their methods of finding oil with them to the midcontinent region. While prospecting in Kansas, Oklahoma, and Texas, Tom occasionally scoffed at geologists who claimed that they could find oil with science, but he did not ignore geologists forever.[12]

8. Tom B. Slick. Courtesy of Western History Collections, University of Oklahoma Libraries.

During the first two decades of the twentieth century, petroleum geologists and their ideas for finding oil gradually gained more credibility in the eyes of some practical men. Slick eventually proved more willing to listen to geologists' advice, but he never invested more authority in their opinions than in his own judgment. He occasionally hired geologists as consultants but ignored their advice when it contradicted his own personal hunches.[13] For example, in 1922 he hired three geologists to survey a tract of land on Laura Endicott's farm in Oklahoma's Kay and Noble counties before investing the money to drill a well.[14] Even though none of the geologists recommended the site, Slick decided to ignore their advice, follow his instincts, and sink a well.[15] The well he drilled eventually produced 4,560 barrels per day and along with the production from neighboring wells initiated the Tonkawa oil boom, a notable chapter in the history of the oil industry.[16] Slick's decision to rely on instinct underscored how successful oil strikes at times resulted from subjective evaluations despite geologists' objective evidence to the contrary. Discoveries like Slick's undermined geologists' claims that they had crafted a scientific approach to oil prospecting. Yet, as demonstrated in Pennsylvania cemeteries, no clear line separated creekology and geology, and the best practical men and geologists realized that the methods of each often complimented those of the other.

Many of Tom Slick's contemporaries considered him one of the luckiest oil finders of his day, but he developed his hunches for locating oil from observations that he made of a region's surface and subsurface geology. Like other practitioners of tacit knowledge, Slick developed a knack or feel for finding oil that grew out of his experiences surveying the topography of Kansas, Oklahoma, and northern Texas.[17] As he explained it, "I know all this country, every foot of it. As a leaser I drove and walked over all of it, studied it, have learned to sort of sense, by intuition, where there ought to be oil."[18] He could not quite explain this knack, believing that he could "sort of sense, by intuition" where to drill. Slick's technique did not rely upon superstition or magic but

9. Tom B. Slick watching oil drain from a storage tank in Oklahoma City, Oklahoma, circa 1927. Courtesy of Western History Collections, University of Oklahoma Libraries.

on his ability to translate experiences traversing local environments into a vernacular knowledge and to apply that know-how successfully.[19] While Slick did not always use the term "geological" to describe the knowledge he acquired at each potential site, he sought information through firsthand observations in order to develop an educated guess about the best possible place to drill. Slick admitted that he never knew with certainty whether he would find oil, "but I have been at it so long, have studied the lay of the land and the underlying formations so persistently, have drilled so many wells, dry and wet, that I often get a hunch."[20] What Slick considered a hunch in fact grew out of "persistently" studying "the lay of the land" and from trial-and-error drilling that yielded important information about the "underlying formations." Only after such extensive preparation could he "sort of feel that there's oil in a certain spot."[21] Slick argued somewhat defensively that his success resulted from a tested methodology rather than good fortune, explaining that "if I strike oil everyone calls it Tom Slick's luck, but do you call that luck? I call it largely judgment based upon experience."[22] Observations of the environment imbued the judgment of many practical men with significant authority in the early days of oil booms on the southern plains.

Slick's son recalled specifically that his father did not reject petroleum geology but that the field had only begun to emerge and that his father adopted the science as it progressed. Forty years after discovery of the Cushing oil field, Slick Jr. recalled that his father "did believe in using whatever scientific method he could, in spite of the rumors that he hired the best geologists available so that he could drill in the areas they condemned."[23] Far from rejecting geology, Slick Sr. was "very interested in the technical phases of the oil industry" even though he lacked "a formal scientific education."[24] Confusion over his father's view of geology struck Slick Jr. so forcibly that he felt the need to set the record straight on another occasion. Although Slick Sr. "was accused by a lot of people of not believing in geology," in fact he was "of quite a scientific frame of mind and . . . believed in it as far as it had gone in those days, but geology hadn't gone very far." Slick Jr. illustrates that practical men like his father succeeded long before petroleum geology coalesced. Perhaps many believed Slick Sr. rejected geology because he "didn't have much geology to go on in those days; the geology wasn't of much value." Without formal principles to guide him, Slick Sr. busied himself "following shoals in wells in the neighborhood, studying how the creeks ran, and various symptoms of that type that you might say were a crude type of geology."[25] Slick Sr. demonstrated clearly the success practical men enjoyed before petroleum geology coalesced as a discipline, and he further showed that they understood geological principles even if they did not concern themselves with publishing their findings in scientific journals.

WESTWARD MIGRATION

Although practical men experienced significant success with the prospecting methods they had crafted at the oil industry's inception, production eventually began to taper off east of the Mississippi River, and they headed for the midcontinent states of Kansas, Oklahoma, and Texas. In the 1890s and early 1900s oil men from Pennsylvania, West Virginia, Ohio, Indiana, and Illinois transplanted themselves onto the southern plains in significant numbers.[26] The population influx rapidly transformed the small town of Bartlesville, Oklahoma, into the oil industry's unofficial capital just a short time after Roswell Johnson left for the University of Pittsburgh.[27] James Veasey, a lawyer who specialized in oil and gas leases, remembered, "That town was literally flooded with old-time producers who had operated in all of the eastern fields."[28] He recalled that "the older operators from the east were flocking there in droves," men steeped in oil field practices that had originated in Pennsylvania's most famous oil sites, places such as Pithole, Pleasantville, Butler, and Clarion counties and the Bradford and McDonald fields.[29] Some of the prospectors who migrated to the midcontinent region had been drilling along the prolific Oil Creek in Pennsylvania as early as the 1860s. J. S. Sidwell began working for the South Penn Oil Company in Pennsylvania beginning in 1896 and followed the industry westward, working in West Virginia, Ohio, and Kentucky before settling in Oklahoma in 1916.[30] Veasey listed several of the men by name and "could mention at least 100 others."[31] Oil men who "were of the same class" as those moving to Bartlesville also settled in eastern Kansas.[32]

Prospectors who moved into the southern plains region continued to rely on traditions and practices for finding oil that they had cultivated since the industry began. Despite Israel White's 1883 demonstration of the anticlinal theory in West Virginia and his articulation of it in print in 1885, practical oil men continued prospecting on the basis of the belt-line theory, or trendology, as they moved into Kansas and Oklahoma. Veasey's work in oil and gas leases brought him into contact with many prospectors, and he remembered "very definitely" that "operators who came there represented every old field in the east, and they brought with them their settled notions regarding oil and gas occurrences."[33] In describing practical men's "settled notions," Veasey recalled that "they would have no part of geology, but on the contrary were still taking leases and drilling their properties under the old belt-line or trend theory, that is, northeast or southwest production."[34] Alf Landon began working in the oil business in 1912 and never made a decision to lease or drill by making "any reference to the guidance of geologists."[35] Other oil men who worked alongside Landon shared this practice; he recalled that "the operators in the field with which I was connected were not at that time following the guidance of geologists in their leasing and drilling activities."[36] Instead of geology, "the old trend theory of northeast-southwest" guided exploration, "and the theory had considerable following for many years."[37] Writing in 1941, Landon contended that "its psychological effect is still in evidence in selling acreage in a block around a wildcat drilling well."[38] Practical men continued prospecting with the belt-line theory because it produced results, but their contempt for geologists also reinforced their loyalty to traditional prospecting methods.

As practical men carried their theories for finding oil westward, many prospectors retained a contempt for geologists.[39] Petroleum geologist Everette DeGolyer recalled that his profession's "intrusion into the industry was generally resented, often with intense bitter-ness" by practical oil men.[40] Writing in 1923, Ralph Arnold, another petroleum geologist, recalled that "twenty years ago a geologist was just as welcome in a drilling rig as a 'hornet at a garden party.' The oil men were prejudiced against us."[41] Arnold understood their animos-ity, however, because he recognized that many practical men had experienced the sting of fraud perpetrated by some creekologists who claimed to possess supernatural powers or unique gifts for locating oil. The possibility that oil men suffered financial loss after investing in one of the many fraudulent schemes prompted him to ask, "When one sizes up some of the freaks and impostors who have posed as geologists, can we blame the oil men?"[42] As the twentieth century dawned, geologists struggled to establish credibility. Practical men who had successfully found oil had maintained confidence in their approach and had no reason to consider geology as an alternative, especially since many had been burned by confidence men hawking the next surefire method for finding oil.

PETROLEUM GEOLOGISTS

The first generation of university-trained petroleum geologists possessed a self-conscious identity as scientific professionals that prevented them from understanding how practical men found oil without formal training as scientists. Everette DeGolyer frequently thought and wrote about the early history of oil prospecting, and he struggled to understand how practical men found oil frequently even though they had not formally studied geology and sometimes refused to acknowledge its validity.[43] Conceding that they discovered some of the most productive and significant oil fields in the United States, from Pennsylvania to California, DeGolyer observed that "some of the early day prospectors must have really had a nose for oil," but he could not more clearly articulate the reasons for their success.[44] Their ability to find oil without systematically employing geology prompted him to acknowledge somewhat self-consciously that "their prospecting was done without the benefit of geological clergy."[45] Despite such a self-effacing comment, he refused to relinquish his authority over the process of finding oil as a professional scientist. He considered oil exploration fundamentally "a geological enterprise" but also felt that "prospecting was more than just geology and it is this 'more' that I am interested in."[46] His training as a petroleum geologist taught him to look for systematic, scientific principles that consistently led to oil, but this orientation prevented him from understanding how practical men historically had proven adept at cultivating tacit knowledge and employing it successfully.

In his quest to understand how laymen's methods comprised "more than just geology," DeGolyer consulted his friend Wallace Pratt, a notable petroleum geologist in his own right, and the two agreed that perseverance bore much of the responsibility for lay success. Pratt considered oil prospectors quintessential American pioneers.

He felt that "the prime requisite to success in oil-finding is freedom to explore," which American culture offered in abundance.[47] Any successful prospector exhibited "the adventurous, chance-taking spirit of the pioneer which pervades America and has impelled Americans to drill thousands of wells every year in search for oil."[48] With this conception in mind, Pratt wrote to DeGolyer that persistence was an "attribute of the successful oil-finder that few of us [petroleum geologists] possess."[49] DeGolyer concurred, observing that all "any of these men really possessed was a great willingness to venture."[50] Most wildcatters felt compelled to persevere in their searches for oil in order to demonstrate the validity of their hunches. They possessed the instincts of an explorer, an unrelenting desire to confirm their suspicions that oil lay in a particular place. Biographers of famous oil men such as Ernest W. Marland, Haroldson L. Hunt, and Tom Slick all attributed their subjects' success in part to the willingness to trust their instincts and to persevere in their searches. Such accounts characterized wildcatters as tough-minded optimists who continued "to hang on, keep on going, and never give up. Slick was of that kind."[51] Perseverance certainly increased practical oil men's chances for success, but without a hunch they had no idea where to begin searching. These hunches constituted a distinct form of knowledge, tacit in character, that prevented petroleum geologists from identifying its form.

Although practical men placed great trust in their instincts, they proved no more successful than geologists at explaining why their hunches consistently proved correct. Edward R. Wilson, a practical oil man based in Oklahoma City, offers another example "of that kind" of prospector who found oil because of the compelling desire to demonstrate the validity of his hunch. He discovered the Pioneer field in Texas even after others abandoned it "due to his unfailing belief in his own intuition and his 'sticktoitiveness.'"[52] When he solicited financial backing from an oil company in Tulsa, "they waved him aside."[53] Wilson's discovery of the field only after "larger companies and big operators were pulling out of the territory" seemed only to confirm the legitimacy of his technique.[54] Although Wilson believed that his instincts led him to oil, he could not articulate any more clearly than DeGolyer why or how this approach worked. On the one hand, he exuded confidence in his approach, declaring emphatically that "I never lost when I played a hunch."[55] On the other hand, he also admitted that even though "my hunch was compelling, almost over-powering, I had no idea that it would lead me to a 15,000-bbl. gusher."[56] Neither practical men nor petroleum geologists understood what constituted a hunch because this prospecting method defied rationalization.

Practical men derived their prospecting techniques from a vernacular form of knowledge that grew out of local experience rather than a scientific understanding of universal earth processes they could systematically employ in different environments. Loosely interpreted as cunning intelligence, the Greek concept metis refers more broadly to an array of practical skills and a specific type of intelligence cultivated in response to social and environmental change.[57] For example, when The Farmer's Almanac suggested planting corn after a specified date, the most cunning farmer adjusted this advice to suit the unique circumstances of his immediate surroundings. The date to plant varied for crops growing at different altitudes and latitudes, in valleys and on hills, and near the coast or inland.[58] The shrewdest farmers adapted the Almanac's universal advice to suit local conditions. Practitioners of such local knowledge did not seek a universal principle based upon rational thought but, instead, applied rules of thumb in their endeavors and invoked a strategy based upon feel, knack, or common sense.[59] Similarly, practical men did not reduce their prospecting methods to deductive principles or codify them into a formal set of procedures based upon rational decisions. Intuition contributed so greatly to their knowledge that they often could not characterize their prospecting methods.[60] This explains Edward Wilson's difficulty justifying the great trust he placed in his instincts or how they led him to oil. Like Wilson, other practical men remained confused over their methodology.

Inability to explain the practical man's success suggested to DeGolyer and Pratt that prospecting skills were qualitatively inferior to the knowledge professional geologists possessed. DeGoyler proposed that many wildcatters simply denied that luck enabled them to find oil and that they offered a rationale for their success even when none existed. To present himself as an authority, a wildcatter was "likely to rationalize his motives" after finding oil particularly when "the reasons for drilling [were] somewhat vague or mistaken."[61] The wildcatter did not consciously lie, DeGolyer conceded, but "honestly believes his revised and entirely fictitious reasoning."[62] To claim authority as an oil-finding expert, the wildcatter offered an explanation that either "fit with then current methods of prospecting" or insisted "mysteriously on a superior knowledge which he does not reveal."[63] Pratt also derided practical men's expertise after conducting a study of oil discovery methods prior to 1911, finding that "ambiguity shrouds records of methods of discoveries. I am convinced that even the fellow who drills a well is sometimes unable to state accurately what his reasons were!"[64] Practical men encountered difficulty articulating their methods because of the implicit experiential nature of their knowledge.[65] Like any experienced practitioner of a particular skill or craft, they developed a repertoire of visual judgments and sensations for assessing their work born out of experience that defied articulation.[66] Successful oil finds that petroleum geologists considered in retrospect as fictitious, lucky, or mysterious actually resulted from local knowledge cultivated through firsthand experience.

Petroleum geologists set out to replace what they considered guesswork in oil finding with a systematic methodology a prospector could apply in a variety of different environments. Investors who equated a hunch to a mere guess worried about the high costs of drilling a dry well and wanted more definitive evidence before investing in a speculative venture. Petroleum geologists courted speculators' confidence by arguing that their prospecting methods employed rational thought rather than intuition, contending that "there's a reason why oil and gas occur and accumulate in certain rocks and regions" and that they espoused privileged knowledge of this reason.[67] By conducting "scientific study," they tried "to eliminate the expensive drilling of dry holes and promote the testing of 'reasonable' looking areas — these men are Oil Geologists."[68] Unlike practical men, geologists strove to articulate a rationale for drilling at a particular site that they believed offered more security to investors. A geologist's reason for thinking a region might produce oil "amounts to more than a 'hunch,' and when he completes a Detailed Survey his opinion is worth big financial backing."[69] Petroleum geologists prospected for oil in a more systematic fashion than practical men, but the attributes of different environments at times resisted the application of generic rules.

LOCAL KNOWLEDGE

Petroleum geologists who conceptualized the principles of their discipline too dogmatically failed to appreciate how knowledge cultivated within a particular locale might yield more practical benefits than universal principles. Western cultures dating back to ancient Greece distinguished between technical knowledge as a type of learning that consisted of hard-and-fast rules, principles, and propositions and intuitive knowledge that depended upon local contexts.[70] Technical knowledge was universal and could be organized into explicit, logical steps.[71] The universal quality meant that such knowledge could be taught as a formal discipline, like geology for instance, whereas practitioners of metis acquired their knowledge through local practice and hands-on experience.[72] For example, the different types of knowledge a riverboat pilot relies upon to navigate a river illustrates the differences between different ways of knowing nature.[73] Every riverboat pilot possesses universal knowledge about rivers and the methods for negotiating the currents, shallows, and turns. However, he also acquires more specific knowledge from his experience on a particular river. Although he can anticipate that seasonal changes will similarly alter water levels in all rivers, only through extended experience can he gain an understanding of how depths will vary at different times of the year on particular stretches of a given river or in certain harbors. The pilot's local knowledge is more practical and therefore superior to the universal rules of navigation. As with riverboat pilots, extended experience in a particular environment often served as better preparation for oil prospectors than did universal rules or principles.

Although a university-trained petroleum geologist inculcated with the technical expertise of a formal discipline, DeGolyer understood that arbitrary factors often undermined universal learning. His undying faith in geology led him to repeat Pratt's often-quoted remark that "the enterprise of winning oil from the earth is essentially a geological venture."[74] Although convinced of his discipline's utility for finding oil, DeGolyer also believed that "geology is not the whole of prospecting."[75] After conducting a study of past oil discoveries, he concluded that they had often been "controlled by arbitraries."[76] Unexpected contingencies such as an unconformity in the environment potentially undermined the application of generic prospecting methods. In the case of oil prospectors, intuitive knowledge at times offered more practical advantages than universal theories. Thus, intuition aided oil prospectors in anticipating arbitraries, or peculiarities of local environments, like those DeGolyer mentioned. Reliance on intuition appeared haphazard to those who witnessed it, so they attributed successful oil finds to sheer luck.

Whether a geologist or a wildcatter, all oil prospectors relied upon luck to some degree, but applying the term too loosely potentially obscured a much more complicated process. DeGolyer believed that "success in exploration depends upon luck and skill," but he hastened to add that "what the proper proportion of each may be, I do not know."[77] Given a choice between the two, he preferred luck over skill. Well sites "selected by the most refined and exact of techniques" could result in failure, and a well drilled "at random for mistaken reasons or no reason at all may result in the discovery of a new and important field."[78] Good fortune improved one's chances for success in any endeavor, but he cautioned that in oil prospecting "one must recognize luck but not overemphasize it"; he warned that the term "is merely a convenient catchall."[79] When petroleum geologists accounted for a successful oil find solely on the basis of luck, "we ascribe to chance the favorable outcome of a complex of conditions, all of which we have not yet been able to analyze, much less understand."[80] DeGolyer's inability to grasp the "complex of conditions" that enabled practical men to find oil underscores the contingent, highly variable quality of the knowledge such men produced and his appreciation for geology's limitations.

As much as DeGolyer strove to objectively evaluate practical men's methods, he felt strongly enough about the inadequacy of these methods to distinguish them from those of petroleum geologists. DeGolyer recalled that geologists' "intrusion into the industry was generally resented, often with intense bitterness" because they "were regarded by most so-called practical oil men as being highly theoretical."[81] Practical men felt that their prospecting methods provided more tangible benefits and were more practical than the arcane theories geologists proposed. DeGolyer reacted defensively to this suggestion and set the record straight by indicating that "the truth is that the practical men were just as theoretical as the geologists and less soundly so."[82] He argued that geologists conducted field work more systematically and therefore more reliably than practical men who encountered the environment haphazardly because they had no formal training in geological theories or how to observe these theories at work in the field. Practical men arrived at generalizations based upon a rather "narrow range of facts which had happened to come within their experience and which were not studied systematically."[83] Unlike petroleum geologists who formulated theories consistent with knowledge they had learned in the classroom and with what they observed outdoors, practical men's "theorizing was uncurbed by any knowledge of the laws governing earth processes."[84] DeGolyer's contention that finding oil was "more than just geology" suggested sympathy with practical men's methods, but as a professionally trained petroleum geologist, he could never entirely shed his orientation that "the winning of petroleum is a geological enterprise."[85]

What geologists perceived as flaws in the prospecting methods of practical men in fact marked the very reasons for the latter's success. Both conducted field work, but their approaches differed. Generally, a geologist gathered information in a regimented and structured fashion and saw himself as "a trained observer, and subjected his theorizing to the limitations of the laws of stratigraphic and structural processes."[86] Unlike practical men, the geologist "based his generalizations on a systematic study of a much wider range of facts."[87] Practical oil men wanted only to solve concrete problems in local environments and did not concern themselves with contributing to a wider body of knowledge.[88] Scientists often denigrated practical knowledge that was too local or context specific and did not lend itself to scientific discourse. DeGolyer echoed this sentiment when he called practical men unsystematic and less theoretically sound and when he alleged that they sampled facts endemic to a particular locale. What he and fellow scientists missed, however, was that the power of local knowledge lay in its variable and contextual character. The conclusions that riverboat pilots, peasant farmers, or practical oil men reached bore greatly on their material well-being and therefore led them to scrutinize the environment more closely and to pay greater attention to local conditions than research scientists.[89] Scientific researchers did not necessarily bear the consequences of their own advice, but the marginal economic status many practitioners of metis held provided an even greater impetus to close careful observation. Living in the field throughout the seasons gave them an advantage in conducting field work because they could observe changing conditions a research scientist might never notice.

CUSHING OIL FIELD

Tom Slick put his close and astute observations into practice in 1912 when he located a favorable site in northeast Oklahoma and drilled the discovery well for the Cushing oil field. Shortly after arriving on the southern plains, Slick departed for Illinois, where he went to work leasing land for Charles B. Shaffer, who had previously become a millionaire in western Pennsylvania's oil fields. Slick prospected for Shaffer

10. Tom B. Slick on the front porch of a hunting camp, probably in St. Charles Bay, Texas, circa 1927. Courtesy of Western History Collections, University of Oklahoma Libraries.

in Kentucky and Canada before the two decided to try Oklahoma. From 1907 to 1911 Slick drilled as many as ten dry holes on land he had leased for Shaffer.[90] Steeped in the tradition of Pennsylvania's practical oil men, Slick relied upon intuition but also actively and consciously evaluated surface and subsurface geological phenomena when deciding where to sink a well. Just prior to his discovery of the Cushing field, he drilled the Tiger well three miles to the east, which, although dry, provided him with valuable information that buoyed his confidence in finding oil nearby. The Tiger well failed to yield commercial quantities of oil, but his drill pulled up positive indications from two thousand feet below the surface. He studied the dips and slopes of the region's surface geology, and these observations led him to Frank Wheeler's farm, where he drilled the Cushing well.[91] Slick considered his decision to drill a hunch, but observable and objective data also influenced his choice. Shaffer and Slick's luck changed in

March 1912 when a well they had been drilling for almost two months came in as a gusher and marked the discovery of the Cushing oil field.[92] Although Shaffer, Slick, and their business partners had leased much of the land surrounding the site, news of Cushing's production brought prospectors and speculators from numerous other states pouring into the surrounding area to repeat Slick's success.

A pivotal moment in the history of the oil industry occurred when Cushing operators began drilling to depths beyond which most practical men believed oil existed.[93] For nine months after Slick and Shaffer drilled the discovery well, "drilling of the shallow sands proceeded rapidly," but operators only realized the field's giant size "when production was found in the Bartlesville sand at 2500 feet."[94] Development of the field proceeded apace, and by August 1913 the total geographic area encompassed an expanse nine miles long and three miles wide.[95] In addition to expanding horizontally, the field also grew vertically as drillers continued finding oil the deeper they went.[96] Eventually six successive layers of sand yielded oil, ranging in depth from one to three thousand feet.[97] The significance of deeper wells lay only partially in the fact that they yielded more oil. The greater depth of Cushing's wells provided geologists with important information they built upon to map the subsurface geology and eventually to find even more oil.

By using the information drillers generated, geologists mapped Cushing's subsurface geology and illustrated in vivid form the principle that Israel White demonstrated in West Virginia almost thirty years before, that a relationship existed between the accumulation of oil and geological structures. Slick found oil at Cushing because he unknowingly drilled into a giant anticline that ran fifteen miles north-south and two to five miles east-west.[98] Similarly, when the Gypsy Oil Company sank a well to the north, it "was found to have been located by chance on a surface anticline."[99] These anticlines, inadvertently located, formed merely the tip of a much larger iceberg. The Cushing field produced significant amounts of oil because it consisted of a number of anticlines clustered together rather than just a single geological structure.[100] Geologists scored a major victory when they illustrated the relationship between these structures and the location of oil.

Even though Slick drilled the initial well, geologists played a pivotal role in the huge amount of oil the field eventually yielded. Although steeped in the prospecting traditions of practical men, Slick had proven willing to consider geologists' recommendations shortly after arriving in Oklahoma. Once drilling began on the Cushing discovery well, he hired geological consultant Lon Lewis Hutchison to report on the prospect.[101] Hutchison soon performed additional consulting work on the Cushing field for another customer, the McMan Oil Company, and in February 1913 he located a structure geologically related to Cushing known as the Dropright dome.[102] The location of this prospect resulted in some of McMan's most valuable production.[103] Everette DeGolyer considered Hutchinson's work pivotal: "I am inclined to think that the first real useful geological work done was the location by Hutchinson of the Dropright dome at Cushing."[104] Several other geologists played important roles in various oil ventures on or near the Cushing anticline, and collectively their work caused many practical men who had once viewed geology with contempt to alter their opinions.[105] With the help of geologists, practical men recognized that they had a new and fairly reliable prospecting method.

Discovery and development of the Cushing field proved a pivotal moment in the history of the oil industry because geologists demonstrated how practical men could reliably and consistently find oil through the application of petroleum geology, specifically using the anticlinal theory. Although identifying a single date or reason practical men accepted petroleum geology risks oversimplifying a much longer and complicated process, most geologists agreed that their profession gained much credibility because of the events at Cushing. The work they performed made the development of this oil field "one of the most important chapters in the history of petroleum geology."[106] By mapping the extent of the anticline and the subsidiary traps in which oil accumulated, petroleum geologists demonstrated that the search for geological structures offered a useful method for finding oil. As James O. Lewis recalled, "Geology was first generally accepted in the Mid-Continent when extensions were successfully predicted by geologists for the Cushing pool."[107] In addition to mapping the outer limits of the anticline, geologists predicted the location of subsidiary traps, or extensions, by applying the anticlinal theory, thus locating additional oil with a greater degree of reliability. DeGolyer agreed that even though "Cushing was discovered in 1912 without benefit of geologic clergy ... it is my opinion that this was probably one of the most important steps toward reduction of the theory to actual working practice. I know of no earlier clean cut application of the anticlinal theory."[108] By demonstrating that they could find oil by applying the theory, geologists had proven their utility to many in the oil industry.

The entry of geologists into the oil industry did not mean that practical men simply abandoned methods that had served them well in the past. Many prospectors ignored geologists and remained skeptical of this emerging science. The most skilled oil men had no reason to take formal geological principles seriously because they continued finding oil with methods that had proven successful long before geologists came onto the scene. Practical men knew, even if only intuitively, that they encountered the environment and cultivated knowledge somewhat differently from geologists, as had their fathers before them, and they trusted themselves to follow the landscape's lead. Although the strongest hunch or intuition could sometimes result in a dry hole, practical men had every reason to remain confident because geologists' predictions also at times failed to produce oil. When geologists began to prove with consistency that their methods worked, some embraced their newly acquired authority by behaving arrogantly and condescendingly toward lifelong practical men who possessed a different cultural conception of oil-yielding landscapes. Some practical men took notice when university boys with scientific methods produced results, but a geologist who failed to find oil after charging a consulting fee remained just another college-educated know-it-all.

4

Institutional Authority

As the twentieth century began, practical men and geologists followed the oil frontier to the southern plains states of Kansas, Oklahoma, and Texas. Throughout the first two decades of the century, practical men and geologists discovered a series of dramatic gushers that unleashed millions of barrels of oil that had been accumulating over millennia. In 1901 prospectors in Beaumont, Texas, followed surface indications of oil and drilled deep enough for the Spindletop gusher to erupt in dramatic fashion, spraying oil for days until they capped the well. This episode repeated itself throughout these states, placing the southern plains region at the center of world oil production. Practical oil men discovered many of these gushers and continued as a presence in the industry but increasingly negotiated for control of the fields with petroleum geologists whose status and authority increased the more they demonstrated their ability to find and control these gushers. Nearly thirty years and half a continent separated Israel White's demonstration of the anticline theory in West Virginia and geologists' mapping of the Cushing anticline in Oklahoma, but many nongeologists within the oil industry remained skeptical of geological science. Their attitudes began to change as more geologists organized the knowledge they produced.

Petroleum geologists acquired more authority throughout the first two decades of the twentieth century by participating in various institutions that provided them opportunities to formulate knowledge that appealed to private industry. Historians typically date the launching of modern American science at approximately the middle of the nineteenth century when the process of organizing knowledge grew into a collective enterprise.[1] Science and industry entered into a reciprocal partnership at this time in which industry benefited from trustworthy advice and scientists found steady and sometimes highly remunerative employment conducting research on problems that interested them.[2] In the oil industry scientists worked on geological surveys and as independent consultants. The relationship between science and industry had grown closer. By the early twentieth century geologists and engineers overcame remaining doubts within the oil industry as businessmen asked not *whether* they wanted geological knowledge but *how* and in what form.[3]

The story of geological science's entry into the oil industry unfolded *not* as a bipolar narrative in which prospectors transitioned from informal to formal ways of crafting knowledge about the environment.[4] Knowledge production occurred not in a linear progression but in a tangled dialectic of private and public institutions in which geologists interacted with each other, the environment, and the constituents who wanted their information. This chapter argues that from approximately 1890 to 1920 petroleum geologists created, influenced, and affiliated themselves with institutions that functioned as venues for fashioning knowledge the oil industry wanted and simultaneously garnered authority and power as professional experts. Prospecting still required physically working in nature, but the process of finding oil reliably demanded more theoretical, arcane, and specialized work practiced by geologists through field work and sanctioned by institutions that included universities as well as state and federal surveys. What constituted valid knowledge began to originate more from geologists' articulation of processes and theories informed by systematic field work than from encounters with nature in which knowledge retained tacit and experiential qualities.[5] Theories formulated on the basis of field work and published in professional journals underwent scrutiny by a growing community of peers. The knowledge of oil accumulation unfolded collectively and collaboratively and was marked by the internal dynamics of an esoteric discipline.[6] Enterprising geologists recognized that the specialized brand of knowledge they practiced complimented the oil industry's need for authority that was hierarchically and bureaucratically organized; they parlayed their expertise into financial gain.[7] Geologists did not necessarily surrender autonomy over the knowledge they provided to oil company managers, administrators, and bureaucrats who paid their salaries. Rather, they maintained control of that knowledge by "internalizing authority" within institutions in which they practiced.[8] These institutions and the environments where geologists practiced functioned together as centers of knowledge production that shaped the education of geologists as an ongoing collective endeavor at the center of economic and political processes. A new culture of professionalism was firmly in place by the second decade of the twentieth century, but the existence of a professional community by no means indicated consensus among petroleum geologists on the scientific principles of oil accumulation.[9] Rather, the increasing stakes for economic and professional power and authority that accompanied growth of the oil industry ensured that the knowledge they generated remained highly contested.

CHARLES N. GOULD

Charles N. Gould recognized the importance of building powerful institutions to advance the field of petroleum geology. He wrote in his memoirs that he pursued three professional goals as a geologist: to hold a chair in geology at a state university, to serve as a state geologist, and to become *the* authority on the geology of the southern plains.[10] The first two goals he achieved within ten years of graduating from college, and the third objective, "to know as thoroughly as one might know" the geology of the Southwest, "has occupied my best endeavors for more than forty years."[11] Gould began cultivating authority

11. Dr. Charles N. Gould, September 16, 1900. Courtesy of Western History Collections, University of Oklahoma Libraries.

as a professional geologist in 1900 when he arrived at the University of Oklahoma to create a geology department. Seven years later, he successfully lobbied for creation of the Oklahoma Geological Survey and assumed its directorship. In 1916 he convened petroleum geologists in Norman, Oklahoma, to inaugurate their discipline's first professional association, the American Association of Petroleum Geologists. What most distinguished his career was his effort to utilize institutional power to garner public, political, and industry support for geologists as arbiters of knowledge for developing oil resources. Gould's career demonstrated that "to know" geology was a collective and ongoing process that required geologists to organize, access, and create authority for knowledge through public and private institutions.

Gould arrived at the University of Oklahoma in 1900 only ten years after its creation, and the sparsely populated campus with few buildings presented itself as the opportunity of a lifetime to a young professor. At the turn of the century, ambitious professionals within the academic world pursued their ambitions at individual, departmental, and institutional levels.[12] The aspirations of the people who sought to advance at any of these three levels did not necessarily coincide, but a strategist who could identify objectives that satisfied goals at each level generated great institutional support. Gould's effort to build support for geology embodied that strategy. At the time of his arrival, the newness of many western universities distinguished them from older colleges to the east that had developed over a longer period

12. Dr. Charles N. Gould with a group of geology students, 1906. Note the hammer in Gould's hand, the compass held by the man in the back row, and two students' geological specimen bags. Courtesy of Western History Collections, University of Oklahoma Libraries.

of time. One or two solitary buildings jutting skyward from the flat southern plains conveyed the sense of institutions imposed upon the landscape by willful people determined to overcome environmental limitations.[13] Gould was among that generation of willful people who saw the university as an institution that facilitated his professional goals, and while he seized the opportunity before him, he also confronted an obstacle in the negative perceptions people held of his beloved discipline. Students held geology in low esteem because they believed it had no practical application.

Gould painted a bleak picture of his first days on campus. Upon arrival "there was no geological equipment whatever, no laboratories, no collections, no books, no students, not even a class room."[14] Only sixty students attended the university, and there was no geology department.[15] Students did not enroll in geology classes because they did not feel the discipline prepared them for future employment: "As

13. First geological field party in Oklahoma camping in Gypsum Hills of Blaine County, August 1900. Courtesy of Western History Collections, University of Oklahoma Libraries.

I remember now, there was no mad rush among the student body to avail themselves of the opportunities offered to secure a first-class geological education."[16] Students did not consider geology one of the "legitimate and recognized methods of securing a permanent meal ticket."[17] Students still considered geology "one of the so-called cultural subjects taught in college. It was a pure science, meaning one that had no known practical application."[18] The perception that geology lacked any practical application had been gradually changing for decades as the hard-rock mining industry recognized its utility and began hiring geology graduates from Stanford University and the University of California.[19] Gould spent much of his career demonstrating geology's practical application to Oklahoma's residents, and the state's enormous reserves of oil greatly assisted him in this effort.

To build institutional support and achieve his professional objectives, Gould spent much of his time publicizing Oklahoma's abundant store of natural resources. One of his first acts upon arriving at the university was to organize a field party to survey the state's terrain and document its natural resources. The university's first official field party departed June 1900 from Norman in a covered wagon consisting of Gould, a botanist, a biologist, and his childhood friend as cook.[20] Heading north from Norman, the party explored central and northwestern Oklahoma. While its members hoped to gather information pertaining to the state's geology, plants, and animals, "especial attention will be paid to the probabilities of gas, oil and coal in certain portions of the territory."[21] The party gathered fossils, minerals, and rock specimens for display at a museum on campus, a practice Gould utilized frequently to publicize the state's resources. For many years he busied himself "preaching the glad gospel of Oklahoma's undeveloped mineral resources" in order to persuade people of the state's "dormant possibilities."[22] In 1908 he published a map that showed oil and gas development throughout the state and explained that "we may expect to find considerable pools of oil" in other locales throughout the state.[23] Gould adopted a more measured tone when addressing fellow geologists on the subject of Oklahoma's future oil production, but still, the future looked promising. He considered it too early to make "an accurate prediction" about future oil production but could say that less than one-tenth of the state's potential production had been extracted. Again, there was "no definite way of estimating" the number of future wells, but he expressed "confidently" that prospectors would find wells "for another fifty years."[24] He interpreted Oklahoma's economic geography as "a case of arrested development," overcome by the state's "energetic people" who transformed their state's "plentiful resources into a prosperous commonwealth."[25] Mineral resources of all kinds abounded, but so much oil flowed that its limit "is not yet in sight," and no one could predict "how much longer this flood of black gold will be poured forth."[26] Gould was utterly and thoroughly a booster who worked for Oklahoma's economic prosperity. His geological expertise served that purpose, and he lost no opportunity to use his credentials to attract the attention of investors to the state.

Although Gould advertised the state's resources remarkably well,

14. Gould regularly assembled booths at the state fair to advertise Oklahoma's resources. He returned to direct the Oklahoma Geological Survey (OGS) after working as a consultant, and he probably prompted this collaborative effort between the OGS and the USGS at the International Petroleum Exposition in Tulsa, September 24 to October 1, 1927. Courtesy of Western History Collections, University of Oklahoma Libraries.

he was not necessarily the most innovative or creative oil prospector. By most accounts Gould was an important geologist who performed pioneering work in the field, but his most important ideas did not relate to petroleum geology.[27] Despite geologists' articulation of the anticlinal theory's relationship to oil several times throughout the nineteenth century and White's demonstration of its utility in 1885, Gould remained unaware that these geological structures potentially housed oil for much of his early career. He explained that he stumbled upon an anticline in 1902 near Newkirk, Oklahoma, but "oil geology was in its infancy and at that time I perhaps did not recognize the importance of the discovery."[28] Everette DeGolyer, perhaps Gould's most famous student, recalled that when he attended the university from 1904 to 1908, "Gould used to take the attitude with us students that 'if we had taken up geology at

15. Oklahoma Geological Survey exhibit at the state fair of Oklahoma, circa 1909. The banner boasts the future yield of various minerals in tons but calculates petroleum potential as unknown. Courtesy of Western History Collections, University of Oklahoma Libraries.

an earlier date, there might have been an opportunity for us in oil geology but that the future did not look very bright as the big fields, such as Spindletop, Glenn Pool, and Caddo, had already been discovered.'"[29] Gould suggested that he became aware of the anticlinal theory's utility in approximately 1911 when he resigned from the survey to work as a private geological consultant. He had prematurely discouraged his students from studying petroleum geology but soon recognized his mistake and decided to use his geological expertise for economic gain. Fortunately for DeGolyer, he did not follow Gould's advice, began working for an oil company, and discovered huge reserves of oil in Mexico. Although still in his early twenties, DeGolyer's income quickly eclipsed Gould's salary as a university professor. Other students who studied under Gould also proved their mentor wrong by parlaying their classroom lessons and field work experiences into employment with oil companies or by starting their own private companies.

In order to generate students' enthusiasm for geology, Gould advocated field work with a sense of mission, but he originally intended the experience to educate students in the principles of geology rather than to enrich them as oil prospectors. He explained that he and other geologists had been "in the thick of the fight for some years" to convey that "the only practical geological laboratory is the field."[30] Classroom instruction simply failed to replicate nature's complexity and thereby proved an inferior venue for teaching students. No teachers, however brilliant, bound to "indoor laboratories" could convey to geology students "something of what it is all about."[31] Students certainly could learn geological principles in lecture halls, but "only by means of field work are geologists made" where the pupil "wears out shoe leather on the rocks" trying to understand his subject.[32]

Gould practiced what he preached by taking students on field trips to various locations, particularly the Arbuckle Mountains in south-central Oklahoma. In 1901 he arrived in the mountains with a wagon carrying his first group of students.[33] Within a few years more than one hundred students had undertaken the pilgrimage.[34] As the department grew, ten to twenty women and three times as many men made the trip annually.[35] Field trips to the Arbuckles continued long after Gould left the university, and he estimated that toward the end of his life as many as five thousand students had visited the mountains.[36] Gould presented field work to students as a form of intellectual work that generated knowledge professional geologists valued as a form of power even if it did not translate into economic gain.

In addition to providing geologists with unique opportunities to study their discipline, the Arbuckle Mountains taught students to appreciate the landscape's aesthetic beauty from a geological perspective. Gould considered the Arbuckles "one of the best places in the United States to study rocks first hand" because the mountains possessed "all sorts and conditions of geological phenomena."[37] The

16. Geology field trip at the Vines Spring Camp in the Arbuckle Mountains, Oklahoma, October 1909. Courtesy of Western History Collections, University of Oklahoma Libraries.

scenery also presented a stimulating and exciting aesthetic experience for those who witnessed it. Gould described the breathtaking beauty of vistas "where one may stand on a hill and see spread out before him in panorama a series of anticlines and synclines — Appalachian-type structure in miniature."[38] This observation shows that anticlines impressed him for their scenic beauty early in his career rather than for their economic value.

On field trips students could see how dynamic geological processes turned rocks on their edges, formed crevices and faults, resulting in "great upheavals of the earth's crust."[39] Gould lapsed into rapture when he described what opportunities the mountains presented: "And the fossils! Where can the American paleontologist find better collecting than here?"[40] White Mound, a famous collecting site composed of white shale and approximately twenty feet high, "stands out on the prairie, and its surface is literally covered with small fossils. One might collect for a year on an area not much larger than a city block and still not pick it clean, for each rain washes out more new forms."[41] Many of Gould's female students gathered fossils at White Mound and later found employment in oil companies as paleontologists correlating

17. Students sketching geological vistas on a field trip to the Arbuckle Mountains, Oklahoma, 1909. Courtesy of Western History Collections, University of Oklahoma Libraries.

layers of stratigraphy in order to map subsurface geology.[42] Experiences of work and play blended into and reinforced one another, and oil companies would soon pay dearly for the knowledge gained from these excursions.

Students' physical and intellectual encounters with nature in the Arbuckles enhanced their aesthetic appreciation of the site and inspired some to choose geology as their profession. Throughout their professional lives, many of Gould's students recalled their field trips as formative experiences. Gould recollected that his students "received their first geological inspiration on these trips."[43] Students who had enrolled in an introductory geology course often chose the discipline as their major course of study after a field trip to the mountains. Many alumni who distinguished themselves as geologists, teachers, and businessmen later in life pointed to their field work in the Arbuckle Mountains as "an eye-opener" and credit their "geological zeal to the camping trip in these mountains."[44] Gould could name at least one hundred prominent geologists who, had it not been for their field experience in the Arbuckles, "would in all probability today have

18. Mostly female geology students studying geological strata on a field trip at Burning Mountain in the Arbuckle Mountains, Oklahoma, 1910. Courtesy of Western History Collections, University of Oklahoma Libraries.

been shoe salesmen or automobile mechanics."[45] Nature educated students in the principles of geology at the same time it taught them something about themselves.

Field work similarly affected students who attended other universities and prompted them to choose geology as their profession and specifically to work in the oil industry. Professor William A. Tarr taught the first summer field work course at the University of Missouri in 1915 at a site near Breckenridge, Colorado. The following summer Professor Edwin B. Branson took the university's geology students to conduct field work near Wind River, Wyoming. Many of the students on these trips eventually worked as chief geologists or exploration managers for major oil companies such as The Houston Company, Carter Oil, and Sinclair. They considered these encounters with nature "rugged work and valuable experience" and acknowledged the importance of these trips as professional preparation because it "changed some of us from students to practical geologists."[46] The field trips taught students to translate their experience into private industry, but they also affected students on a deeper, more personal level.

Field trips to geological sites facilitated geologists' attempts to build

19. Women geology students gathering small fossils at White Mound in the Arbuckle Mountains, Oklahoma, 1909. Courtesy of Western History Collections, University of Oklahoma Libraries.

institutional authority within the universities and surveys in which they worked. Gould recognized that field work served his professional goals because the excursions excited students about the discipline of geology, effectively training them in the field, increasing the number of geology majors, and building a pool of student labor he drew upon to pursue goals of the state survey when he became its director. Gould never intended and did not foresee that these experiences would lend themselves to oil prospecting, but the oil industry began hiring many of his students in large numbers when the ability to identify, interpret, and map geological structures such as anticlines morphed from an aesthetic experience into the most important method of finding oil. Field work conducted by university students in Oklahoma during the first two decades of the twentieth century offers an example of how cultural perceptions of nature within a specific time and place defied categorization as solely scientific or recreational. The knowledge they produced at times reflected personal and individual experiences but also grew from their collective and collaborative efforts. Geologists who succeeded most frequently in the oil industry proved adept at synthesizing multiple intellectual and physical experiences and transforming fieldwork into an interpretation of nature that explained how and where oil resided.

Geological interpretations remained open to debate, and geologists cultivated authority by competing with one another for knowing nature best and for publicly claiming credit for discoveries of oil at the expense of other geologists with whom they had collaborated. The opportunities for professional advancement and economic gain prompted Gould to seize upon opportunities for discovering a significant oil field. He and his ex-student and brother-in-law, Everett Carpenter, worked together as oil prospectors for Empire Gas and Fuel Company.

Carpenter recounted his early professional experiences to fellow geologist Edgar Owen, who was preparing to write a history of petroleum geology, but he explained that he "would like no comment about" his relationship with Gould in the book.[47] The two men had very different recollections of who discovered the Augusta and El Dorado oil fields in Kansas, important for their abundant production that enabled Empire to expand operations as a major producer throughout Kansas, Oklahoma, and Texas. Carpenter further recalled that Gould initially negotiated terms for a contract with Empire's vice president Alfred Diescher to work as a private consultant but subsequently "ignored" Diescher and finalized contractual details with officials in the company's Pittsburgh headquarters. Alienated by this action, Diescher "would never again receive Gould." The contract committed Gould to find oil for the company, but "he tried to get the leasing in his own hands," which was ethically questionable. Carpenter gave Gould the benefit of a doubt, stating that "whether he had an ulterior motive in mind I do not know but I doubt it." Diescher suspected that someone in the company leaked proprietary information when another geologist, Charles H. Taylor, Gould's personal friend and university colleague, secured nearby leases and formed his own oil company. Carpenter described Gould as "a hound for publicity" and that he "liked to see his name on the front page" of newspapers. After he and Carpenter had finished their collaboration, an Oklahoma City newspaper quoted Gould as saying that he had discovered a site in Kansas where oil would soon be found.[48]

Gould's apparent scheming to acquire oil leases at the same time he received consulting fees and Carpenter's recollection of his brother-in-law as a publicity hound reveal how some petroleum geologists sought authority by publicizing their scientific efforts and by hoping to parlay their reputation into financial gain. Carpenter's account provides a useful window into the coupling between science and industry as it existed in the early twentieth century and hints that this marriage did not always constitute a match made in heaven. Gould and Carpenter understood their relationships to private industry in very different ways. Both men embraced opportunities for steady employment increasingly available to geologists within the oil industry. As the next chapter will show, Carpenter eventually directed Empire Fuel and Oil Company's geological research department, which consisted of more than two hundred geologists. Gould never worked full-time on a company payroll but resigned his position at the university and geological survey to work as a private consultant for several oil companies. The career choices of the two geologists reveal that Carpenter demonstrated a stronger willingness to balance corporate interests with his own personal, professional, and economic ambitions than did Gould. Scientists and the oil industry forged alliances during the first two decades of the twentieth century, but these relationships took many forms as geologists pursued their interests and renegotiated boundaries between themselves and private and public institutions that hired them. Personal and professional motivations for economic power continued to shape the configurations of pure and applied science.

For all of Gould's success as a university professor, his failure to profit from the discovery of oil while many of his students grew wealthy may have left him an embittered man. Gould certainly prospected for oil, but he discovered "a number of dusters" in areas he "firmly believed" oil resided.[49] B. W. Beebe, a geologist who knew Gould and discovered his unpublished memoirs, noted that the document lacked any mention of Gould's theoretical research and its later economic importance. Gould failed to mention his discovery of an enormous buried mountain range that extended from Oklahoma into Texas that later geologists explained accounted for significant supplies of oil and gas.[50] When Gould was asked why he did not publicize the contribution of his theoretical research to later oil discoveries, his only response was "I hate oil geology."[51] According to Beebe, Gould disliked "the money side of the profession," and while he reaped "adequate" financial rewards, "wealth eluded him."[52] Gould's disillusionment with petroleum geology grew when sites he had recommended yielded oil but at depths deeper than he had drilled, making "a sad experience" for a man whose work generated wealth for others but not for himself.[53] Gould's research produced many important geological insights, but he failed to forge a relationship with industry as successfully as other geologists.

In the western states a paradigm shift occurred among geologists who began to realize that anticlines accounted for much of the oil in their environments. Geology had been applied in the search for oil both formally and informally since the oil industry began in the United States. Widespread acceptance of the anticlinal theory, however, marked an important turning point in the relationship between science and industry. The ability of geologists to locate anticlinal structures and to predict oil's location with increasing frequency convinced skeptical laymen of geology's utility and presented geologists with a clear and reliable prospecting method. Historian of science Thomas S. Kuhn has written that when scientists experience a gestalt in their thinking, their perceptions of the natural world change. Revolutions in thought enable scientists to "see new and different things" even when they look in familiar places. For example, a person who could only perceive the exterior of a box could conceptualize the box's interior from below after experiencing a gestalt. A student looking at a contour map sees lines on paper, but training in cartography allows him to envision terrain in three-dimensional form.[54] In short, "when paradigms change the world itself changes with them."[55] The scientist's altered perception allows him to enter a new world not fixed by nature on the one hand and science on the other but a world "determined jointly" by the environment and the scientific tradition in which he has been trained.[56] The growing recognition and acceptance of the anticlinal theory by geologists and laymen transformed the oil industry as Kuhn has described by altering prospectors' perceptions of environment. Petroleum geologists began looking at familiar landscapes and seeing anticlines where previously they had seen none.

As popularity of the anticlinal theory increased within the geological community, Gould experienced a Kuhnian paradigm shift of his own and began to perceive familiar environments with keener scientific insight. He first noticed anticlines in the Texas Panhandle when he traveled there in 1903 as a private consultant for the Reclamation Service, who hired him to search for water.[57] While riding horseback along the Canadian River twenty miles north of Amarillo, "I had noticed some outcrops of dolomite ledges among the red beds, and observed that, in some places, these ledges were dipping at rather high angles, suggesting the presence of an anticline."[58] On a similar expedition two years later, Gould "located several anticlines, or domes," which he described in the report he wrote for the United States Geological Survey (USGS).[59] He attributed no significance to the surface geology he observed because "at this time there was no particular interest attached to anticlines. The relation between geological structure and oil and gas accumulation was suspected by a few geologists, but it was by no means a matter of common knowledge."[60] Within approximately five years, more geologists associated anticlines with oil accumulation and transformed what had been a theory into the dominant prospecting method in most oil-rich states. Gould's perspective on anticlines shifted from "no particular interest" to great enthusiasm and precipitated his resignation from the geological survey he had created in order to prospect for oil full-time.

When a wholesale grocer from Amarillo walked into Gould's geological consulting office, he hoped to find a tract of oil-rich land in Oklahoma and only considered a location near his home in the Texas Panhandle as an afterthought. Millard C. Nobles hired Gould in 1916 for advice about finding oil on land in Tishomingo, Oklahoma. When Gould reported unfavorably on the prospect, Nobles asked if he knew of any oil or gas prospects near Amarillo. Initially Gould stated that he did not "but happened to remember the anticlines along the Canadian north of Amarillo. So I told him what I had seen when doing work for the government thirteen years before."[61] The new paradigm in Gould's head prompted him to reconsider the relationship between anticlinal structures he had previously observed. Nobles immediately expressed interest in the prospect, organized the Amarillo Oil Company, and hired Gould to survey the area. Ten businessmen who made up the company each invested one thousand dollars, and by following Gould's advice, they later sold the company for one million dollars. As news spread of their success, nearby ranchers hired Gould to survey their property and his consulting company eventually mapped nine different geological structures in the area.[62] For three years prior to this initial discovery in the Texas Panhandle, geologists in Oklahoma had been demonstrating that they could find oil by locating and mapping anticlines, and investors throughout the region had begun to take notice.

Within the first two decades of the twentieth century, more geologists used their ability to identify and map anticlines as an opportunity to elevate their professional standing. Petroleum geology coalesced as a discipline within a very short period of time, and the geologists who entered the profession from approximately 1900 to 1920 recognized these years as pivotal because the anticlinal theory dominated their thinking. Dorsey Hager, a geologist well known within the oil industry at the time, argued that anticlines mattered greatly in finding oil in Kansas and Oklahoma. He found that of seventy-five productive oil sites discovered in the two states from 1915 to 1916, all but four existed on "well-defined structure" and sixty percent of the total were on anticlinal domes.[63] Jesse V. Howell remembered that from 1910 to 1915 "the anticlinal theory enjoyed high standing in the scientific world, and acquired a prestige" within the oil industry.[64] Harold B. Goodrich, well known among the first generation of petroleum geologists, remembered 1911 to 1921 as a pivotal period with "a very complicated history" but as a time of "progress."[65] He paid homage to Israel C. White as "our guide and the pioneer of our profession."[66] Led by this pioneer, practitioners of the new science multiplied in number. Goodrich recalled that in 1911 "there were not many geologists in Oklahoma" but that the number grew to "probably five hundred" in ten years.[67] Numerical growth coincided with a proliferation of new ideas. Goodrich cast the period in triumphal terms, stating "the main point that impresses one is that the entire oil business is one of constant progress."[68] Petroleum geologists embraced their popularity among oil producers and rejoiced in their elevated authority, which they considered part of the industry's "constant progress." The geologist, as DeGolyer argued, constituted "the most important tool" in that progress because he "possessed in his mind" perceptions and viewpoints for discovering oil. The narrative of pioneering progress that geologists fashioned, in which their innovative minds led industry out of darkness and into light, fails to capture the contested nature of geological knowledge and its origin within the complicated matrix of the institutions in which they worked.

UNITED STATES GEOLOGICAL SURVEY

The federal government provided a training ground for geologists and taught them the same skills Gould's students learned through field work. Prior to 1900, although the federal government had begun to organize or administer geology, it only indirectly aided petroleum exploration by training geologists as cartographers. Multiple government-sponsored surveys of the American West existed after the Civil War, but creation of the USGS consolidated these efforts in 1879.[69] Charles D. Walcott, who directed the survey from 1894 to 1907, supervised nine reports on oil production, but the federal survey did little else to aid specifically in oil exploration.[70] Survey geologists spent much of their time making topographical maps that Director John Wesley Powell planned to combine into a national atlas documenting the country's natural resources.[71] Western states welcomed efforts to map their resources, and their elected officials protested whenever budget cuts prevented USGS geologists from making them.[72] The topographical skills geologists acquired working at the survey contributed to the transformation taking place within the oil industry.

Petroleum geologists acknowledged the central role the USGS played in preparing them for their profession. Carroll H. Wegemann graduated from the University of Wisconsin with a degree in anatomy and studied geology for two years at the graduate level before joining the USGS in 1907. Survey personnel supervised him for two years studying coal bed geology as preparation for heading a field party to evaluate coal prospects during his third year. Observing geological structures and processes firsthand enabled Wegemann to make better maps. He worked in Salt Creek, Wyoming, under a noted survey geologist who "personally inspected my work and gave me many valuable suggestions in field mapping."[73] As his confidence grew, he increased the quality of his maps. He characterized his expertise at this learned art with nonchalance: "It was merely a matter of very careful mapping of the surface beds in order to work out the structure."[74] Wegemann's humility belied the superior skill required in the use of the plane table and telescopic alidade, instruments necessary for topographical mapping. In addition to the good training he received at the survey, his supervisor designed and gave him one of the first small-scale telescopic alidades. Wegemann expanded upon his training and developed "methods of my own for its rapid use."[75] Training by senior geologists within the survey imparted expertise, but the best geologists cultivated knowledge by drawing upon their intellects and instincts once alone in the field.

20. The first topographical field party in the Arbuckle Mountains, 1905. The plane table, alidade, and stadia rod were becoming standard geological instruments for making topographical maps. One of Charles Gould's most noted students, Everette DeGolyer, is at the plane table sketching. Key Wolf is holding the stadia rod. Courtesy of Western History Collections, University of Oklahoma Libraries.

The unique opportunities the USGS provided for observation and mapping prepared geologists to work in the oil industry better than studying geological principles in classrooms. Geologists who worked for the USGS practiced making topographical maps more than any other activity, and their expertise captured the oil industry's attention. One geologist recalled that "nearly all the early work of petroleum geologists for the oil companies was the mapping of surface structure."[76] Making maps required "a certain aptitude but only a very elementary training in geology."[77] Although Wegemann conceded that a petroleum geologist should possess an education in the sciences that included chemistry, physics, and geology, "he is likely to learn more in one season as assistant to a good geologist than in two years of schooling." Field work educated geologists in a manner classroom lectures could not, for "in geology it is field experience which counts more than college training." Working for the USGS transformed Wegemann's life. He felt as though he had "specialized in oil geology" due to his survey work, "each new assignment adding to my experience and knowledge." He considered himself a pioneer in the work he performed for the government, "aiding in the development of a science which . . . was not taught in any school."[78] Survey geologists quickly parlayed their map-making skills into higher-paying jobs in private industry, revealing an increased awareness among businessmen of geology's utility.

California's practical men accepted geology due in large part to the efforts of the USGS. The survey significantly influenced prospectors in the state to accept geology as a guide in locating oil when the USGS published a number of studies from approximately 1900 to 1910.[79] In 1901 a survey geologist named George Homans Eldridge began investigating and documenting every occurrence of oil in California, and the survey published his findings in 1907.[80] Other survey publications followed and provided a foundation for further study by later geologists.[81] These bulletins convinced many practical oil men of geology's utility in their wildcatting ventures, and within a short time the major California enterprises such as the Kern, Associated, Union, and Standard companies employed geologists full-time.[82] In the words of Carroll Wegemann, "Someone had to do the original pioneer work to establish in the field the facts of petroleum geology. That work was done in this country first by the men of the USGS for the most part, but later by these same men in the employ of the oil companies or by men whom they trained."[83] Geologists received their training as topographers in the public sector and parlayed this experience into better-paying jobs in the private sector. This pattern repeated itself in oil-rich states throughout the nation.

Under Powell the USGS used many of its topographers to locate and map geology relating to water resources, but its priorities began to shift in the 1890s under Walcott, who fostered a closer relationship between mining companies and geology. Government-sponsored science and private industry continued growing closer together into the twentieth century when Congress appropriated more money for the survey to investigate fuel resources including coal, lignite, and petroleum.[84]

21. The University of Oklahoma geology department and United States Geological Survey men, 1910. Gould is the fourth seated man from the left with his legs crossed. Courtesy of Western History Collections, University of Oklahoma Libraries.

The survey hired specialists in mining geology and technology who branched off in 1910 to form a new institution, the Bureau of Mines. The discipline of geology was growing more specialized in order to serve constituents in private industry whose interests became more closely intertwined with geological expertise. Directors of state geological surveys also wanted to serve private industry with the knowledge they gathered but lacked the large budget of the federal government. The USGS cooperated with state surveys by deploying its geologists to make topographical maps of their resources.

OKLAHOMA GEOLOGICAL SURVEY

Prior to Oklahoma statehood in 1907, Gould relied upon his keen ability to access institutional power to coax the USGS into mapping large portions of the Oklahoma and Indian territories. The professional contacts Gould made early in his career working with geologists and administrators at the USGS provided valuable field experience but also taught him how to access the resources of this federal agency. The USGS had not always collaborated and coordinated its work with individual states. By 1900, however, the national and state surveys cooperated more closely, and Gould worked throughout his career to utilize the USGS whenever possible to acquire their assistance in mapping.[85] The topographical branch of the survey prepared maps for large portions of Missouri, Kansas, and Texas.[86] Gould made one of his first geological field trips with Robert T. Hill, a prominent member of the USGS, while a young man living in Kansas.[87] The federal survey employed Gould in 1901 when one of its geologists, Joseph A. Taff, who had spent years mapping the coal fields of Indian Territory, hired him to assist a field party.[88] Gould again worked for the USGS when the Reclamation Service hired him to survey and map the region's water resources. He traveled to Fort Worth, Texas, where he met the bureau's director, Frederick Newell, to discuss the goals of his employment. Gould recruited four geology students to accompany him, and in 1903 they drove a covered wagon throughout western Oklahoma, northeastern New Mexico, and northern Texas.[89] Gould used the data gathered during this and subsequent expeditions to write three USGS water-supply publications.[90] Despite these early and frequent contacts with the federal survey, he encountered difficulty getting the agency's assistance to investigate Oklahoma's mineral resources.

Having witnessed the power of federal agencies to facilitate western resource development, Gould traveled to Washington to court politicians and to request that the USGS dispatch geologists to assist him with topographical mapping. Without federal assistance Gould struggled to produce topographical maps in high volume and quality. Prompted by the knowledge that the USGS cooperated with many state surveys and by the "large amounts of money" Congress appropriated to this end, Gould sought federal assistance but met with "considerable difficulty" obtaining funds.[91] After a letter-writing campaign failed to produce results, he sensed that Oklahoma stood "in bad repute" among government officials.[92] Ever the astute politi-

22. Gould rejoined the Oklahoma Geological Survey later in his career and probably prompted this 1927 exhibit at the International Petroleum Exposition in Tulsa. The placard on the wall reads, "Oklahoma Geological Survey in cooperation with Tulsa geologists and the U.S. Geological Survey." Note the mineral samples, maps of Oklahoma, and the prominent role of the surveying equipment. Courtesy of Western History Collections, University of Oklahoma Libraries.

cian, Gould adopted a more direct approach and "having failed to get results by correspondence, I made a special trip to Washington to interview the authorities of the U.S. Geological Survey."[93] The strategy succeeded, and he began to receive funding. Gould recruited local politicians to his cause as well, namely Oklahoma's Senator Robert L. Owen, who proved "instrumental in helping to secure this aid."[94] These experiences reinforced a lesson Gould had been learning since he arrived in Oklahoma, that institutional power could unleash or restrict geological expertise where matters of resource development were concerned. His lobbying efforts paid dividends, and "the next year it was not so difficult to get co-operation."[95] Eventually the USGS prepared topographical maps for more than one-half of Oklahoma.[96]

These results reinforced in Gould's mind the political nature of his chosen profession.

The same political savvy Gould employed to acquire federal assistance he used to influence creation and funding of the Oklahoma Geological Survey (OGS). Due to his professional ambition, foresight, and shrewd political instincts, Gould almost single-handedly brought about creation of the state survey in 1907. When the territorial legislature had met the previous year to discuss the possibility of statehood, he identified an opportunity to realize his long-held professional goal of becoming a state geologist. The perfect moment had arrived, for he "long had in mind the establishment of a geological survey in the new state."[97] In addition to calculating the right time to broach the idea, he drew upon personal ties to ensure that the survey came to fruition. He cultivated personal relationships with several members of the state constitutional convention and "had little difficulty in having appointed a committee on a geological survey."[98] Once the convention agreed to consider mandating a geological survey in the state constitution, Gould found his way into the discussions and personally guided the debate. Although not an official delegate to the convention, "I met with the committee several times and aided them in formulating plans for the establishment of a survey."[99] Due in large part to his lobbying and guidance, Oklahoma became the only state with a provision in its constitution mandating that the legislature establish a geological survey.

Not satisfied with merely a constitutional mandate, Gould took the liberty of drafting a law to ensure that the state organized and implemented a survey in the manner he had envisioned. As soon as the delegates adopted the constitution, he wasted no time formulating this law "for carrying out the plans I had in mind for so many years."[100] He wrote to various states and countries to obtain copies of the laws governing their surveys and sought advice from other state geologists. He aimed at writing a law that was both concise and "flexible," not burdened "with too many provisions."[101] He hoped this approach would keep the survey free from political entanglements.

The law that Gould drafted avoided the "danger" of a large governing board unable to compromise when faced with difficult decisions.[102] He refused to leave the appointment of the state geologist in the hands of any single individual, "not even the governor of the state," because he wanted to ensure that whoever served as director remained free of political constraints.[103]

The final version of the law he drafted codified the University of Oklahoma's influence over the survey and left the state geologist unencumbered by any legal restrictions. Gould's bill called for the creation of a Geological Survey Commission to consist of the governor, the state superintendent of public instruction, and the president of the University of Oklahoma. These three officials appointed the director of the survey, who, because of Gould's careful wording, "can do almost anything he desires."[104] Once again Gould relied upon personal suasion to ensure that the legislature passed his bill. Content that he had written "the kind of law I wanted," he campaigned for its passage, and "again my acquaintance over the state stood me in good stead."[105] Because he knew personally almost half of the state legislature and many of the constituents of those legislators with whom he was not familiar, the bill faced little opposition, and the governor signed it into law in 1908. In his memoirs Gould admitted no pretensions to serving as director of the survey, but he should not have been too surprised when the governor asked him to fill the position since Gould had so visibly and aggressively lobbied for the bill's passage.

Because of Gould's dual role as founder of the university's geology department and the state's geological survey, the line between the two entities often blurred, and at times no line appeared to exist at all. In addition to vesting the university president with authority to select the director, Gould further blurred distinctions between the two institutions by stating in his bill that the University of Oklahoma should furnish rooms and equipment until the survey could provide them. Just before the legislature passed the bill, however, a campus fire eliminated extra space at the university, and Gould rented four rooms near his residence to house the OGS. Located so close to his home, Gould had nearly unlimited oversight. With Gould in charge of both the university geology department and the geological survey, little distinction, if any, existed between the two. For example, the university catalog for 1913 included more than four pages detailing the nature of the survey's work and publications produced, and the catalog further explained to curious students the advantages of studying at an institution that housed the state's geological research headquarters. The geology department and survey's activities continued to overlap even after Gould resigned to work in the private sector. Daniel W. Ohern assumed the directorship and also chaired the geology department, which gave him access to students.

The students Gould and Ohern hired conducted field work for the OGS and in the process gained experience that prepared them to make significant contributions to the discipline of petroleum geology as private consultants and as employees of major oil companies. In the opinion of petroleum geologist Jesse Howell, "The Oklahoma Geological Survey had a decisive part in the sudden progress of petroleum geology during and immediately following the Cushing boom, not so much by its publications as by the professional work of the geologists whom it trained."[106] The combination of course work taken at the university, field trips to locations around the state, and field experience working for the OGS taught students how to systematically prospect for oil through the conscious application of geological principles, which increasingly set them apart from prospectors who relied upon less formal techniques. The joint efforts of the geology department and geological survey trained a generation of geologists whose efforts began to transform how prospectors searched for oil.

Gould wasted little time placing field work at the center of the OGS's activities. He had never organized a survey before, but his professional instincts guided him clearly and decisively. He confessed having only "an inkling of what was to be done" to organize a survey, but his decisive action indicated otherwise: "Within an hour after

23. Daniel W. Ohern succeeded Gould as director of the Oklahoma Geological Survey and directed this geological expedition on the Verdigris River, 1909. Included in this picture are Ohern, Charles N. Gould, Everett Carpenter, Arthur Reeds, Ben Bolt, and Robert Wood. Courtesy of Western History Collections, University of Oklahoma Libraries.

my appointment I had long distance telephone calls to five men in different parts of Oklahoma, directing them to begin field work."[107] By the end of the summer, he had arranged for nine different field parties to conduct work throughout Oklahoma.[108] Gould found a ready supply of assistants to hire in the geology classes he taught at the university. Lon Lewis Hutchison studied at the University of Oklahoma and parlayed the knowledge he gained into a profitable business as an independent consultant before most oil companies proved willing to hire petroleum geologists. While at the university, he enrolled in classes such as economic geology, paleontology, and mineralogy.[109] After graduating, he attended Yale University, where he received a master of science degree.[110] In addition to his academic training in geology, he also gained experience observing firsthand the principles he had learned in the classroom. While attending college in Oklahoma, he worked for the Oklahoma Geological Survey and studied the stratigraphy of oil fields while traveling in a covered wagon in the northeastern corner of the state.[111] His astute ability to observe geological structures and processes in the field prompted Gould to comment that Hutchison was "one of the best men I have ever taken

24. Daniel W. Ohern's field party having lunch, 1910. This picture includes Ohern, Charles N. Gould, Everett Carpenter, Artie Reeds, Ben Bolt, and Bob Wood. Courtesy of Western History Collections, University of Oklahoma Libraries.

out. He has a keen eye, correlates accurately and is able to express intelligently what he has observed."[112] After graduating from Yale in 1908, Hutchison returned to Oklahoma, where he accepted a position as assistant director of the OGS under Gould and headed a field party that surveyed the northeastern counties he had traveled as a student.[113] Hutchison quit the Oklahoma Geological Survey after two years and moved to Tulsa, where he opened an office as a geological consultant.

MAPPING CUSHING'S ANTICLINES

The discovery of a relationship between anticlines and oil in Cushing, Oklahoma, in 1913 accelerated demand for petroleum geologists by practical men and oil companies who increasingly recognized the utility of their services.[114] Hutchison found his consulting services in such high demand that he became "the first man in the state to devote his entire time to petroleum geology," suggesting that a new era in the history of the oil industry and the place of geologists within it had begun.[115] Hutchison's university and survey experiences enabled him to increase the probability of finding oil before investors incurred the expense of drilling. His consulting fees increased so much that by 1911 he realized he could afford to resign from the OGS because "my private affair matters bring me a living."[116] Although careful never to charge for advice he gave while working for the OGS, Hutchison had written a number of reports for individuals and companies, including the practical oil man Tom Slick, who had drilled the initial well at Cushing. Hutchison maximized production of oil at this field by applying geology to locate additional well sites. Several other geologists participated in various oil ventures on or near the Cushing anticline, and their work collectively caused many practical men who had once viewed geology with contempt to begin viewing it favorably.[117]

Like Hutchison, Frank Buttram took geology classes at the University of Oklahoma and worked for the Oklahoma Geological Survey, and his work on the Cushing field further convinced practical men to take petroleum geology seriously. Buttram's personal history and the manner in which he entered the oil industry illustrate how petroleum geologists and practical men began to diverge as disparate groups within the oil industry. Whereas Slick was "born among the oil derricks" in Pennsylvania and smelled oil upon taking his first breath of air, Buttram was born in 1886 to a poor farming family who had migrated to Indian Territory from Missouri.[118] He worked his way through the Teachers' Institute at Tecumseh, Oklahoma, received high grades, and earned a certificate to teach school. After finding employment as a teacher, he continued taking college classes at the Normal School in Edmond until 1909. By this time he had acquired enough credits to enter the University of Oklahoma as a junior from which he received a bachelor's degree in chemistry.[119] Buttram continued his education, and while pursuing a master's degree, he caught the attention of Gould, who hired him to work as a chemist for the survey.[120] This position familiarized Buttram with the state's geology and initiated his work in the oil business.[121] He authored three bulletins while working for the OGS, and one, "The Cushing Oil Field," functioned as a gestalt that changed the way prospectors found oil and how the oil industry viewed geologists.[122]

Buttram's bulletin and its accompanying maps demonstrated to practical men that a relationship existed between anticlines and the accumulation of oil. Approximately four months after Slick drilled the Cushing discovery well in March 1913, Buttram led a team of geologists in an effort to survey and map the oil field.[123] The OGS devoted most of its resources that year to this task.[124] In December 1914 the Oklahoma Geological Survey published bulletin number eighteen, a report written by Buttram that was the "best description of any mid-continent oil field which had ever been published."[125] Beyond just describing the field, Buttram's real accomplishment lay in demonstrating that Cushing's subsurface geology bore a direct relationship to the location of oil. By mapping the area's surface and subsurface geology, he convinced practical men that "geologists were able to predict the limits of production of Cushing."[126] His ability to articulate and illustrate geology's relationship to oil made his work "the first important oil publication in Oklahoma."[127] When practical men found oil on the basis of Buttram's maps, geologists "gained the confidence of the oil operators" because they had provided a predictable method for locating oil.[128] With geologists' maps to assist them, "the structure of these fields was taken as the type to be sought elsewhere," and practical men expanded out from the initial well to find additional domes or anticlines in each direction.[129] The more success practical men scored on the basis of information gleaned from Buttram's maps, the more they believed that geology, and particularly the anticlinal theory, could lead them to oil. As they found additional oil throughout 1914 by drilling into deeper sands, "the proof by drilling that geological maps of surface outcrops delimited oil accumulation gave further impetus to geological reconnaissance."[130]

Buttram's work at Cushing placed petroleum geology on a firm foundation and provided impetus for his departure from the state

25. North end of the Cushing oil field along the Cimarron River, Oklahoma. Courtesy of Western History Collections, University of Oklahoma Libraries.

survey and entry into private industry. Buttram's report on the Cushing field caught the attention of executives for the Metropolitan Life Insurance Company, who met with him in their New York offices to ask, "How can you get us into the oil business in Oklahoma?"[131] He left the meeting with a $40,000 advance to finance the Fortuna Oil Company, a venture in which he collaborated with Ohern, who had been his boss and director of the state survey. Both men received stock in the company and quit their jobs working for the state to devote their full attention to building the company. Over the next few years Fortuna demonstrated "an outstanding record of success" based solely upon the recommendations of Buttram and Ohern.[132] Because of their geological expertise, the first seven out of eight wells they drilled resulted in good production.[133] Buttram and Ohern achieved these successes by "the discovery of surface structures," demonstrating the validity of the ideas Buttram and other geologists had been articulating.[134] The company quickly gained a reputation because of its geologists' successful track record in locating oil.[135] As the industry began to take notice of Fortuna's success, the company sold a

forty-eight-acre lease to the Roxana Petroleum Company in 1915 for one million dollars.[136] In less than four years, Fortuna acquired and developed a significant number of oil fields, and the original investors sold out to the Magnolia Oil Company in 1918 for eight million dollars.[137] At thirty-two Buttram was a millionaire. Subsequent to the sale of Fortuna, he started his own oil company, and in the early 1920s he began acquiring large leases near the Corsicana field in east Texas and in Oklahoma's Wewoka-Seminole field.[138] In exchanging their positions as state employees for more remunerative possibilities in the private sector, Buttram and Ohern followed the lead of their mentor Gould, who had also quit the OGS in order to make more money working for the oil industry.

The anticlinal theory initiated a new era for the oil industry because it demonstrated clearly geology's utility to laymen, but the theory soon grew into a dogmatic formula and prompted geologists to call for new ideas that could lead to oil. After years of less systematic approaches to prospecting that included various forms of creekology and application of the belt-line theory, which potentially left investors and speculators vulnerable to swindlers, the oil industry welcomed a clear

26. An Oklahoma Geological Survey exhibit at the state fair of Oklahoma, circa 1912–16. The model portrays a cross-section of an anticline with a derrick on top that conveys petroleum geology to laymen in a simplified manner. Courtesy of Western History Collections, University of Oklahoma Libraries.

and comprehensible explanation of how the earth sciences decreased the probability of drilling dry wells, expending capital needlessly, and increased the probability of finding oil. The theory's usefulness remained contingent upon the environments where it was applied. As early as 1909 geologists who specialized in locating petroleum began decrying its limitations. DeGolyer dubbed the anticlinal theory a "philosophical speculation" that made sense logically but had lost much of its utility because geologists applied it dogmatically and without sufficient observations in the field.[139] Geologist Frederick Clapp wrote frequently about his discipline's economic utility. He considered the term "anticlinal theory" an "unfortunate" naming because it gave "a non-geologist the idea that the adherents to this theory believe the distribution of oil and gas pools to be dependent on anticlinal structure and on nothing else. Such is not true, however." Clapp explained "scores of times" to drillers and oil investors that while he endorsed the theory, it was "only *one* of the factors in the accumulation of oil and gas pools" and that geologists had "to consider every other particle of evidence found" in the structure.[140] Clapp's experiences revealed that laymen and geologists privileged the theory to such an extent that it had grown into a dogmatic belief that failed to account for different environments where oil pooled for geological reasons unrelated to anticlines. Even in environments where anticlines abounded, not all of their folded strata necessarily contained oil. Geologists perfected the practice of locating and mapping anticlines, but drilling still remained the only certain method for finding oil.[141]

Many practical men continued drilling for oil successfully even though the science of petroleum geology developed increasingly under the purview of authorities who trained within universities and geological surveys. Gould's career demonstrates how scientists who cultivated institutional power had a greater capacity to affect the oil industry as the twentieth century began even if as individuals they did not make for the best prospectors. University professors and survey directors had access to large numbers of eager young men and women whom they trained in the rudiments of their discipline and dispatched into private industry, forever altering how oil companies acquired new supplies of oil. Attracting students translated into professional opportunities for university professors and survey directors to enhance their public prestige and expand the authority their institutions wielded. These professors and directors increasingly practiced as members of a more pronounced community of professional petroleum geologists. University students and survey employees parlayed their knowledge into economic gain as private industry recognized the utility and application of their theoretical skills and mapping expertise. Of course, opportunities for students, scientists, and industrialists to prosper depended upon the environments where they searched for oil. Organized geological knowledge made great strides in a few decades, but it did not by itself guarantee success in finding oil. The complexities of subsurface geological environments and their relationship to oil ensured that disputes among prospectors continued and that consensus among scientists and engineers joined in the common cause of finding and producing oil would not come easily.

3

Appropriated Knowledge

5

Geology Organized

HENRY L. DOHERTY'S TECHNOLOGICAL SYSTEM

If Charles Gould used public institutions to professionalize petroleum geologists, Henry L. Doherty exercised private power to advance their professional status by hiring them in his various oil and gas enterprises. Gould's and Doherty's actions demonstrated that the organizational changes occurring in American society throughout the late nineteenth and early twentieth centuries took place in both the private and public sectors.[1] In the private sector business ventures in the chemical, electrical, and oil industries began to organize into large economies of scale. Standard Oil Company epitomized this push for consolidation in its very name and in its business practices, but initially John D. Rockefeller focused the company's efforts on refining, marketing, distribution, and transportation and left the task of locating oil to others. By the early twentieth century the Supreme Court declared Standard Oil an illegal monopoly and broke up the trust, thereby complicating its ability to access the southern plains states, which collectively provided the majority of the world's oil supply.[2] Standard's absence from the region as a monopoly created a vacuum that Doherty and others eagerly filled. Doherty recognized the utility of geologists in systematically prospecting for oil and gas and hired them in large numbers to staff his various oil and gas enterprises. Doherty's act completed the industry's vertical integration that Rockefeller had initiated and won for geologists more authority, legitimacy, and credibility within the industry.

27. This memorial sketch of Henry L. Doherty casts him in heroic iconography as the builder of a vast technological system comprised of oil derricks, refineries, electricity transmission stations, and other infrastructure. Courtesy of Western History Collections, University of Oklahoma Libraries.

Doherty's greatest innovation lay not just in hiring geologists but in his capacity to function as a system-builder.[3] The term "system-builder" characterizes an innovator who possesses the instincts and mentality of an inventor, engineer, industrial scientist, manager, and entrepreneur.[4] Doherty functioned in all these capacities and fashioned in his Cities Service Holding Company a system of corporate entities with petroleum geologists and engineers as integral members used to locate oil as efficiently and expeditiously as possible and to conduct research into the mechanics of petroleum reservoirs in order to maximize their yield. In a technological system an invention functions as more than simply utilitarian hardware but reflects the context of the people and organization who produce it.[5] Thus Doherty created a context for material and intellectual innovations to produce oil that consisted of technology, ideas, and people. He hired university-trained geologists and engineers from nearby universities and the USGS to

28. One of many Cities Service gas stations that represented Henry L. Doherty's effort to consolidate control over the production, distribution, and marketing of oil-based fuel. Courtesy of Western History Collections, University of Oklahoma Libraries.

innovate alongside craftsmen and technicians. Technological systems often possessed regional styles because of the close proximity of natural resources necessary to make the system function.[6] Doherty built his technological system on the southern plains by conjoining oil prospecting and production technology with the region's enormous stores of oil resources. His efforts added legitimacy to the profession of petroleum geology and provided an impetus to the emergence of another new specialty, petroleum engineering, which was aimed at maximizing efficiency.

DEVELOPER OF MEN

Doherty perfectly fit the image of a Horatio Alger hero, and the story of his success so captivated those who heard it that they often attributed Cities Service's accomplishments to him alone, even though many

29. Henry L. Doherty. Courtesy of Western History Collections, University of Oklahoma Libraries.

other employees also deserved credit for the company's innovations. One senior-level employee, Everett Carpenter, expressed awe at his boss's remarkable rise to success: "The story of Henry L. Doherty is a Horatio Alger story that outshines all the Horatio Alger stories at their best."[7] Although the rise of Cities Service owed much to Doherty's engineering genius, he exhibited perhaps even greater brilliance in devising programs to train and educate his work force. He hoped to cultivate innovators like himself who could craft technical solutions to problems the company encountered. Doherty declared, "If I am ever known for anything I would prefer to be known as a man who could develop men rather than a man who could pick men."[8] This image remained fixed in the minds of other Cities Service personnel.

W. Alton Jones, Doherty's successor, remembered him as a man dedicated to recruiting highly educated personnel and working aggressively to inculcate employees who lacked an education with knowledge that would benefit the company. Jones conceded that Doherty's charismatic and towering persona often overshadowed contributions of other key personnel: "Mr. Doherty's eminence was such that many gained an impression of Cities Service as a one-man organization. Nothing could be further from the truth."[9] Far from denigrating Doherty's contributions, however, Jones hailed him as an innovator, particularly because he "devoted his principal attention to the development of men."[10] Doherty pioneered "in the recruitment of engineering talent from the schools and universities," and when the workers he hired lacked sufficient training, he developed "special training schools to instruct these recruits in applying their specialties to the requirements of Cities Service and of the business world."[11] His commitment to recruiting and sustaining a highly educated and well-trained work force guided his efforts to build Cities Service into a dominant holding company of gas and electric utilities.

Doherty grew convinced of the need for university-trained engineers while managing Denver Light and Electric. He felt that the existing work force lacked the kind of knowledge necessary for preparing the company to adapt to changes soon to transform the industry. Hands-on experience provided the best training for workers to learn their jobs since the gas industry began. Self-educated workers continued to dominate the industry into the twentieth century, a point Doherty confirmed in 1904 when he suggested that for the majority of employees in the gas utility business, one "might classify [them] under the greatly abused term of 'practical men.'"[12] Although lacking a technical background, these workers possessed "valuable practical experience" and were "the kind of men that can never be entirely eliminated and there is no desire to do so."[13] Practical men contributed significantly to the industry throughout the nineteenth century, but the necessity of providing gas for new markets in growing cities increasingly required skills they did not possess. To make this point, Doherty cited two examples of superintendents he had known who could neither read nor write but secured "better than average results of all gas companies throughout the country at that time."[14] Although they were "remarkable men in their way," he contended that they "would be entirely out of place with our present methods," not to mention that "the gas business already possesses a preponderance of this class of men."[15] Of course, workers who lacked technical training could always educate themselves by reading company manuals in their spare time as Doherty had done, but most chose not to exercise this option.

Even though Doherty reached a position of eminence within the gas utility business by educating himself, he believed a gradual but drastic transformation had been taking place within the industry that would eventually make all but the most brilliant self-educated workmen obsolete. As America's cities grew, companies looked for technologically sophisticated solutions to the problems of locating and supplying gas to expanding markets. Organizational changes necessary to accommodate new ways of conducting business required skills and knowledge that many workers did not possess. Doherty observed that "the gas business was formerly carried on as a craft, but is gradually undergoing a change and will eventually be carried on as a science."[16]

He placed practical men, who made up the bulk of the work force, squarely within the craft tradition because "they were not scientists, their education was limited, and they were at least indifferent to scientific methods and technical training."[17] Far from remaining indifferent, Doherty embraced science and technology, and this orientation set him apart from the typical practical man.

What bothered him more than practical men's aversion to new ideas was their outright contempt for university-educated engineers who initiated many of the changes he believed the gas industry required. Reflecting upon his statement about practical men's indifference toward science and technology, Doherty observed that "indifference is hardly the word to use" because many expressed outright "opposition to technically trained men and to scientific and exact methods."[18] In fact, he observed, "it is not hard to find men in the gas business who have an outspoken contempt for a college education."[19] Even though he too lacked a formal education, Doherty embraced the university as a training ground for prospective employees. He hoped to recruit a new type of employee, one "thoroughly educated and possessing all the necessary fundamental knowledge to enable them to take up the special problems of the gas business. This, to the writer's mind, means a college educated man."[20] Doherty's commitment to hiring engineers grew so strong that he equated opposition to this idea as an impediment to the industry's growth. Practical men had grown so entrenched in their beliefs about how to conduct their jobs that they blocked the entry of new ideas that might invigorate the industry. In stubbornly refusing to acknowledge when their methods had grown obsolete, practical men "insist in spite of increased knowledge and changed conditions, on following what was considered good practice twenty years ago."[21] They failed to recognize that "conditions are changing and we must change to meet them."[22] Doherty believed that many practical men remained provincially wedded to traditions that had served them in the past and that they were unwilling to consider new ways of conducting their work: "they do not take kindly to proposed changes and prefer to see the business conducted along the lines of traditions of doubtful origin."[23] Clinging to outdated work habits even as the industry changed, "the man who may have been a leader in the gas fraternity twenty years ago may be an obstacle to progress today."[24] Slowly but surely, however, "these obstacles are being overcome."[25] The example of college-educated men who had made "remarkable progress" served as "an incentive for others to endure the direct opposition, and sometimes ridicule, to which they have been compelled to subject themselves."[26] Throughout the first decade of the twentieth century and increasingly thereafter, Doherty hired those engineering graduates bold enough to run the gauntlet of ridicule dished out by the industry's practical men.

Even though some universities began engineering programs during the last two decades of the nineteenth century, Doherty initially could not find graduates with specific knowledge of the gas and electric industry. This lack of university-trained specialists prompted him to create a school designed to supplement the theoretical knowledge of college-educated engineers with practical experience. Due to the shortage of engineers, he initially hired graduates of Denver's technological high school, gave them the title of apprentice, and insisted that they attend an informal instructional program designed to provide on-the-job training.[27] He encouraged the apprentices to continue acquiring theoretical knowledge by enrolling in correspondence schools, offering to pay up to 100 percent of their tuition.[28] Around 1905 Doherty began hiring graduates of engineering colleges from throughout the nation and added a more formal cast to the training program, dubbing it the Doherty School of Practice and granting new recruits the title cadet engineer.[29] Even though instructors employed a prescribed curriculum and taught in a more formal classroom setting, the school maintained its goal of providing cadets with a forum in which they "could obtain practical experience to supplement their theoretical education."[30] From 1904 to 1932 the Doherty training school admitted 1,059 engineers, almost half of whom remained with the company at the end of that period.[31] Although the school successfully infused the company with new ideas that prepared for sweeping changes Doherty believed would soon transform the gas and electric utility industries, the company never lost sight of the need to strike a delicate balance between theoretical information and practical experience.

Although company executives welcomed the opportunity to integrate the latest technological ideas into a work force that had been dominated by practical men, they saw the school as a vehicle to ensure that engineers did not get lost in a theoretical haze and that they learned how to apply their knowledge in order to enhance efficiency and streamline production. Jones described the training program as a forum "to develop men who can apply their technical training with good sense and good judgment."[32] Instructors at the school echoed this sentiment. They strove to indoctrinate engineers with a perspective wide enough to understand how the company functioned but critical enough to identify problems without getting lost in the details of day-to-day operations. Given this orientation, personality played a significant role in the type of men admitted: "the kind of men we need are men with large enough vision to see the problem as a whole and separate the essential object to be achieved from the details now in vogue for achieving these essentials."[33] To ensure that students received every opportunity to acquire practical experience, instructors required them to spend a certain amount of time in each department. This approach worked well, particularly because "the men being fresh from school absorb information rapidly and eagerly grasp the opportunity to supplement their theoretical training with a wide range of practical knowledge."[34] Doherty's School of Practice proved such a success for training young men in the utility industry that he applied the same philosophy to the engineers he hired after acquiring several oil and natural gas companies.

Already convinced that university-educated engineers would play a vital role in the gas and electric utility industries, Doherty acted quickly to hire similar recruits and train them at his School of Practice when

he purchased more than fifty oil and gas companies. In 1912 he organized Empire Gas and Fuel to serve as the parent company of several
subsidiaries he purchased throughout Kansas, Oklahoma, and Arkan-
sas. Just as he had trained engineers to work in his utility companies,
he employed the same strategy to fill the ranks of these companies
with employees who possessed a mix of technological knowledge and
hands-on experience. As the oil and natural gas industries boomed
during the first two decades of the twentieth century, "the demand
for trained men in this work became urgent" for Empire executives.[35]
The company responded by opening an additional training program in
Bartlesville, Oklahoma, "to train technical graduates in the oil and gas
business."[36] In addition to engineers, the Bartlesville program recruited
university geology graduates.[37] The program accepted eight recruits
in 1916 and twenty-five the following year.[38] Although no students
entered the program in 1918, the number skyrocketed to one hundred
in 1919.[39] Numbers tapered off thereafter, probably a reflection of the
oil industry's boom-and-bust cycle.

As with the public utility training course, Doherty's school in oil and
gas production required engineers to spend time in all departments
in order to gain as much practical experience as possible. Students
at the Bartlesville branch participated in tasks as diverse as account-
ing, scouting, auto service, and warehouse work. Of course, they also
spent time in the oil and gas fields, where "they learn production at
first hand, as roustabouts on the leases and clerks in the field offices.
They are thus brought in direct contact with field problems, both
production and engineering."[40] As much as Doherty pinned his hopes
on young engineers to lead Empire Gas and Fuel into the future, the
training program aimed to maximize their exposure to practical men
such as roustabouts and clerks who could provide a perspective engineers would never have received either in a classroom or an office. As
Americans consumed more oil and Empire expanded its operations
to include gasoline manufacturing plants and oil refineries, students
spent time in these facilities too and learned "by actual experience

how the various products of crude oil are made and prepared for the market."[41] Practical men demonstrated by example how to perform specific tasks engineers had never encountered, and in this respect contributed as much to the learning process as did the instructors. The Doherty School of Practice drew upon a wealth of experience by exposing its engineers at every turn to practical men working in the field at tasks many had been performing all their lives.

BARNSDALL AND STRAIGHT

Even before Empire established the Bartleville training program, the work traditions of practical men influenced Empire's approach to oil production because the previous owner of the subsidiaries that made up this parent company built them from a lifetime of experiences as a Pennsylvania oil man. Doherty purchased the fifty-six oil and gas companies that became Empire Gas and Fuel Company in 1912 from Theodore N. Barnsdall, a Pittsburgh-based practical oil man. Empire's three most important subsidiaries were the Wichita Natural Gas Company, the Quapaw Gas Company, and a large interest in the Indian Territory Illuminating Oil Company (ITIO).[42] By acquiring these companies, Doherty marked his entrance into the midcontinent oil and gas industry but also inherited the problem of locating new natural gas reserves, which had been running dangerously low. The men Doherty's company assembled provided a solution to the problem of diminishing natural gas reserves and placed petroleum geology on a firm foundation within the oil and gas industry. One of the men who contributed to this effort was Barnsdall himself, who retained a minority interest in each property and continued to play a role in management-level decisions.[43]

Barnsdall was a second-generation Pennsylvania practical oil man who had achieved great success and wealth by the time he and Doherty came into contact. The success of Drake's well in 1859 inspired Barnsdall's father William to undertake his own search for oil, and he drilled the nation's second commercial well in a nearby location on his

30. Theodore N. Barnsdall. Courtesy of Western History Collections, University of Oklahoma Libraries.

brother-in-law's farm.[44] As Pennsylvania's oil industry boomed, William soon acquired the wealth to send his second son Theodore to a preparatory school, but after a brief stint, he realized that a job in Barnsdall Oil Company rather than a scholar's life better suited the rebellious lad.[45] Theodore thrived in the atmosphere of a booming oil town and seemed at play among the confidence men, wildcatters, speculators, and gamblers.[46] He quickly earned the reputation of a savvy businessman who instinctively recognized a good deal, and at age sixteen he successfully brought in his own well.[47] By 1912 Theodore had extended his holdings to Oklahoma and Kansas when his westernmost companies caught the attention of Doherty.

Barnsdall's aversion to academic life and his instinctive business acumen enabled him to blend well among practical oil men, but these characteristics did not prevent him from hiring William B. Pine, a prospector who systematically studied geology in order to find oil rather than relying upon hunch or intuition. Shortly after 1903 when Pine arrived in Indian Territory, he began studying geology. Prior to accepting this position, he had spent approximately three years working for an oil-well equipment distributor and in his spare time supplemented his income as a creek skimmer.[48] This activity involved skimming the surface of creeks to collect oil that leaked into the water from natural reserves beneath the ground. Pine's work experiences taught him many of the most rudimentary aspects of the early oil industry's production and prospecting techniques. Finding oil through the application of geology

HIS FAVORITE DIVERSIONS
WERE HUNTING AND FISHING.
MEMBER EPISCOPAL CHURCH
AND VARIOUS MASONIC ORDERS.
MARRIED IN 1901 TO KATHARINE
RANDALL HASKELL, HAD THREE
CHILDREN, HARRIET, LOIS AND
RUSSELL.

BOYHOOD AMBITION TO BE A
BIG OIL PRODUCER. BORN
IN TIDIOUTE, PA. ATTENDED
MILITARY ACADEMY ON THE
HUDSON, NEAR WEST POINT.
GRADUATE STANFORD UNIV.,
HONOR STUDENT AND MEMBER
OF THE FOOTBALL TEAM.

HERBERT R. STRAIGHT

HAS BEEN ACKNOWLEDGED AS "DEAN OF
THE MID-CONTINENT PETROLEOCRATS,"
"PIONEER IN PRODUCTION," AND "OUTSTANDING
OILMAN OF OKLAHOMA." HE INITIATED MANY
PIONEERING INNOVATIONS IN EXPLORATION,
PRODUCTION, CONSERVATION AND MANAGE-
MENT GROUNDED ON THE FUSING OF THE
KNOW-HOW AND PRACTICAL KNOWLEDGE OF
THE OILFIELD VETERAN WITH THE TECH-
NICAL THEORY OF THE ENGINEERING
GRADUATE.

HIS FULL-TIME CAREER IN OIL BEGAN AT THE AGE
OF 16 AS A ROUSTABOUT IN THE BRADFORD OIL
FIELDS. HE MOVED TO BARTLESVILLE IN 1911
AND, IN 1912, CONSOLIDATED A GROUP OF
PRODUCING PROPERTIES WHICH WERE TO
BECOME CITIES SERVICE OIL CO. HE ROSE TO
COMPANY PRESIDENT AND BOARD CHAIRMAN,
RETIRING IN 1946. HE CONTINUED HIS
ACTIVE INTEREST IN THE COMPANY AS A
DIRECTOR UNTIL HIS DEATH MAY 4, 1963

31. Herbert Straight embodied the collaborative tradition cultivated and taught within the Cities Service holding company's oil and gas subsidiaries. Straight sought to fuse skilled laborers' practical knowledge with university-educated geologists' and engineers' theoretical training. Courtesy of Western History Collections, University of Oklahoma Libraries.

did not qualify as accepted practice during his informal apprenticeship, but Pine began studying the subject through a correspondence course.[49] In 1906 he parlayed what he had learned into a job scouting oil reserves for Barnsdall.[50] Over the course of three years, Pine worked with Barnsdall's general manager, F. M. Robinson, to acquire forty thousand acres of oil leases.[51] When Barnsdall decided to relinquish control of the leases, Pine and Robinson took them over. Pine earned enough money to initiate wildcat prospects of his own and achieved significant success as an independent operator.[52] He believed that the application of geology lay behind his prosperity, and this belief differentiated him from other independent oil men who continued to rely on intuition, dowsing, and other less scientific methods.[53] Whether Pine's geological prospecting method by itself won him Barnsdall's favor remains unclear. What is clear, however, is that Barnsdall showed no fear in hiring men who used this emerging, new science.

Although motivated primarily to help a friend in need, Barnsdall bucked the traditional antipathy practical men felt for geologists when he hired Herbert R. Straight, who possessed a degree in geology from Stanford University. Other than Straight's university education, the two men had much in common. Straight's father was also among Pennsylvania's early oil pioneers, following on the heels of Drake and William Barnsdall by drilling the nation's third commercial well.[54] Like many practical men who relied on intuition rather than science for finding oil, the elder Straight "was imaginative and creative in his approach to drilling and to prospecting for oil."[55] This expertise directly benefited his son because "during summer vacations, it was customary for Herbert to work on his fathers [sic] leases near Bradford Penna [sic], where his father lived."[56] Herbert took leave from helping his father for three and a half years in order to attend Stanford University, where he graduated with a degree in geology in 1896.[57] After school he returned to Pennsylvania and worked for his father's oil company until 1911 when the elder Straight "suffered financial reverses" and could no longer employ his son.[58]

Herbert soon found employment when his friend Barnsdall offered him a job overseeing oil and gas properties in Oklahoma. When Doherty acquired Barnsdall's holdings, his decision to retain Straight as general manager of the oil and gas division revealed an affinity for scientifically trained experts that was ahead of its time in the oil and gas industries but consistent with the philosophy Doherty had employed since his earliest days in business.[59] Straight's degree in geology qualified him as the new breed of university-educated employee Empire increasingly sought to hire, but in assembling the company Doherty chose executives and managers who complemented each other with a mix of practical experience and theoretical knowledge. As president of the new properties he chose a longtime Pennsylvania oil man, J. C. McDowell, who had worked as general manager of the famous J. M. Guffey Petroleum Company, which had played a significant role in Spindletop and later grew into the Gulf Oil Company.[60] Guffey and his partner John H. Galey earned a reputation as two of the oil industry's most famous wildcatters and two of the earliest to hire geologists.[61] Guffey and Galey greatly facilitated the industry's acceptance of geology in 1885 when they hired Israel C. White, who found oil by applying the anticlinal theory.[62] Not surprisingly, their general manager McDowell also embraced geology, and he hired others who recognized benefits of this emerging new science.

ALFRED J. DIESCHER

Alfred J. Diescher served as Empire's vice president and, like Doherty, had little formal education but acquired technical expertise as a self-taught engineer, producing a number of innovations that benefited the company.[63] The Diescher family was of German descent, but Alfred's father Samuel emigrated from Hungary after studying at Karlsruhe University and the University of Zurich and after working as a mechanical designer for various industrial enterprises throughout Europe.[64] In America the senior Diescher built a successful engineering firm and designed and constructed projects such as steel bridges,

32. Cities Service geological crew, early 1900s. Everett Carpenter is seated and third from the left. Courtesy of Western History Collections, University of Oklahoma Libraries.

incline planes, pipelines, oil pumping stations, and gas compressor stations.[65] Alfred quit high school during his first semester to begin working in his father's Pittsburgh engineering practice and eventually became a full member of the firm.[66] Although he lacked a formal education in engineering, Alfred worked in his spare time studying the field and preparing himself as an engineer. A subordinate of Diescher's at Empire, Everett Carpenter, recalled that his boss "mastered Trigonometry, Analytical Geometry, Differential and Integral Calculus all on his own."[67] Carpenter had received a geology degree from the University of Oklahoma, but his formal education did not diminish the esteem he held for the self-educated Diescher, whom he considered a "clear thinking and far seeing individual with a keenly analytical mind."[68] Diescher embraced the science of geology just as he had embraced engineering. According to one account, Diescher paid "much attention to geology" prior to 1913, the year he hired Carpenter for his geological expertise.[69]

Diescher placed enough faith in geology to hire Carpenter, who he hoped would find additional reserves of natural gas to supply the subsidiaries Doherty had purchased from Barnsdall. The field work Carpenter practiced as a student at the University of Oklahoma and later at the USGS facilitated his ability to locate and map the anticlines that held the gas Diescher desired. In 1913 Carpenter began exploring near Augusta, Kansas, because "some small showings had been found in some wildcat wells."[70] Nothing particularly "geological" led him to the site other than the "small showings," and like some practical men, he merely looked for oil by following the success of others.[71] After arriving in the immediate vicinity, however, his approach changed significantly. He began surveying the area to acquaint himself with the surface geology and identified two stratigraphic layers of limestone that he then "mapped by pacing the distances from numerous points on the outcrops to Section lines and corners."[72] Carpenter drew upon skills he had learned at the USGS to depict in visual form the geology he had observed. After carefully traversing the area, he documented the surface geology so that "the outcrops were appropriately indicated and colored on a map."[73] Rather than merely drawing pictures of what he had observed, he showed "the inclination or dip of the rocks . . . by the conventional dip and strike symbols then in use."[74] With his work completed, Carpenter expressed satisfaction that "it was a noble map — one that would warm the heart of the most cold blooded."[75] With map in hand he set out to convince his boss how he could find natural gas with this visual representation as his guide.

Carpenter encountered a problem when he returned to Empire's Bartlesville office because company executives, although not averse to geologists, could not entirely understand their methods. Carpenter took great pride in his map but lamented that "there was just one thing wrong with it — the general manager could not understand what was meant to be shown."[76] Steeped in the traditions of practical men, both Diescher and Barnsdall proved willing to hire geologists but initially failed to comprehend the significance of the anticline Carpenter depicted because they could not read his map. Carpenter included different perspectives of the structure, explaining that "cross sections were tried but without absolute elevations it was difficult to give them the rosy appearance that graces the modern reports."[77] Diescher could not make sense of the map's graphic symbols, but "still a good enough picture was made to enable me to convince him that a detailed survey with a telescopic alidade was necessary."[78] Paradoxically Diescher approved Carpenter's request for additional survey work but not the funds necessary to purchase the required equipment.

Diescher's willingness to hire geologists but reject some of their techniques demonstrated how persistently some practical men resisted the scientific and technological changes transforming the oil industry. Carpenter began exploring at Augusta with "none of the tools so necessary to a geologist such as a compass, aneroid, hand level, etc., . . . and for transportation I had my own two legs."[79] Although he eventually acquired a compass and hand level, "it was still forbidden to spend any money even for an alidade," and he "did not get permission to buy one" even though Diescher agreed to additional field work.[80] Carpenter "finally succeeded" in overcoming his boss's intransigence but admitted that "it took considerable scheming and a lot of nerve to get around that hurdle."[81] He waited until company officials left the office for a two-week period so that he could place an order for the alidade and receive it before they returned. They eventually uncovered his scheme, and "I was called on the carpet for a rather severe lecture."[82] Carpenter received a tongue-lashing, but he had acquired the tool he needed and initiated the field work necessary for creating an even more detailed map.

Carpenter's second survey produced such a superior map that it convinced Empire executives to lease acreage and begin drilling on the Augusta anticline. Carpenter solicited the services of another University of Oklahoma graduate, J. Russell Crabtree, to carry out the field work. Crabtree studied engineering rather than geology but found his knowledge equipped him to survey and map the area. Crabtree grew into one of Empire's "best men in geological mapping. His maps were works of art."[83] He used an alidade to measure geological structures with more precision and contour lines to represent them more clearly on the map. Crabtree's work greatly impressed Empire executives because "the structure was portrayed by means of contours which the management could visualize and understand at a glance."[84] Carpenter invoked his best public relations skills when he met with Barnsdall and Diescher to show them the map and "to explain the geology of the Augusta field."[85] At the end of their meeting, "Barnsdall asked if it was our contention and belief that simply because the contour lines ran around in complete circles that that made this a good place to drill."[86] Carpenter answered affirmatively, and on that basis Barnsdall "gave his consent to whatever it was that was desired."[87] The use of contour lines on the new map showed more clearly the anticline's elevation and convinced Empire executives to invest in drilling at the site.

AUGUSTA AND EL DORADO

Carpenter's discovery of the Augusta anticline and oil field prompted reconnaissance of the surrounding area and led to the location of an even larger anticline and oil field at El Dorado, Kansas. The two discoveries won acceptance for petroleum geology within the oil industry. Petroleum geologists working at the time of the finds remembered them as pivotal moments for their discipline's acceptance by nongeologists in the oil industry. One geologist declared, "Empire invented Oil Company geology, based on experiences in Augusta and El Dorado, Kansas."[88] Other companies had previously used geologists but mostly as consultants and in a much less systematic way than Empire. Edgar Owen, one of the first geologists Carpenter hired, expressed clearly and succinctly how Empire's approach to geological exploration differed from any previous effort. The company's discovery of these two fields "was most influential in popularizing the application of geology at a time when the Mid-Continent was becoming the most important district in the world."[89] Empire's achievement lay not in originating

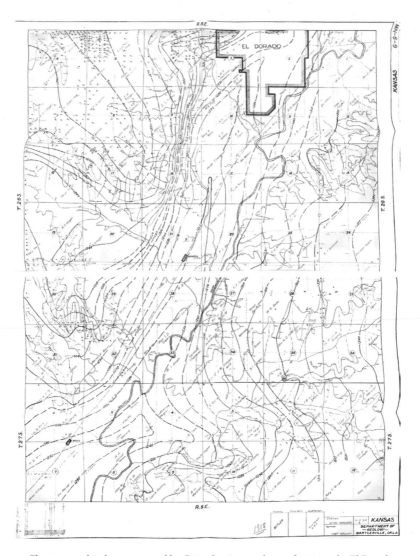

33. The topographical map prepared by Cities Service employees depicting the El Dorado anticline and associated elevations of nearby structures, 1918. Courtesy of Western History Collections, University of Oklahoma Libraries.

any new exploration techniques but in demonstrating "the economic utility of a program consisting of geological reconnaissance followed by detailed mapping of surface anticlines and promptly leasing and drilling of large blocks on the favorable structures."[90] After the success at Augusta and El Dorado, Empire executives no longer needed convincing of the utility in Carpenter's methods and reasoned that if one geologist could find oil, then many geologists could find a lot more.

Confidence in Carpenter's methods ran so high among Empire's top-level managers that they ordered him to hire as many geologists as he could find, and in doing so, they created the largest geological exploration department the oil industry had ever seen. Empire committed significant resources to acquiring a staff that could conduct field work on a large scale and translate observations into high-quality contour maps that outlined the structural characteristics of anticlines. After receiving the order "to increase the personnel of the department greatly before the Standard group and other large companies gathered up all the geologists," Carpenter acted swiftly and began combing the geology departments of nearby universities for as many students and faculty as he could find.[91] He particularly drew upon his close ties at the University of Oklahoma to build his staff. As an instructor there recalled, "Mr. Carpenter employed six or eight of the graduating class from Oklahoma University in June 1914; in fact, practically every man that had majored in geology."[92] A geologist working for the company remembered how swiftly the number of his coworkers grew as "Empire expanded its geological staff from practically nothing at the beginning of 1915 to more than 100 men early in 1916."[93] By the summer of 1917, the company had hired "about 250" geologists and placed them "just about everywhere" throughout the United States, including New York, Kentucky, Texas, Ohio, and Wyoming.[94] Accounts vary as to exactly how many geologists Empire employed, but the number certainly exceeded two hundred and may have approached two hundred and fifty.[95] No longer content to invest in drilling on the basis of a hunch, Empire committed itself to geologists who could identify and map

34. Cities Service geological staff standing in front of its corporate headquarters, 1916. Courtesy of Western History Collections, University of Oklahoma Libraries.

anticlinal surface structures. This commitment increased the probability of finding oil.

A problem quickly surfaced in this plan, however, because the students and faculty Carpenter hired from universities understood geological theories and principles but did not know how to construct contour maps or operate the necessary surveying equipment. Even though professors like Charles Gould thoroughly indoctrinated their students through field work, their excursions tended to focus on examining geological processes rather than practical tasks the oil industry required such as mapping the formations they had observed. The nature of field work students practiced to fulfill university degree requirements differed from that performed by employees of the USGS. Geologists like Carpenter, who had attended the University of Oklahoma and who had married Gould's sister, learned how to make maps and isolate resources while conducting field work for the USGS prior to joining Empire. Although Empire and other oil companies that followed its lead initially "were raiding the faculties" of universities, they eventually began instead to recruit geologists from the USGS. One geologist explained that "the reason that the companies preferred survey men to faculty men was that the latter were on the whole less good field men."[96]

Still, following the lead of Doherty, Empire executives believed that they could teach university geologists the practical skills their company required. The Bartlesville branch of Doherty's School of Practice aimed at teaching university-educated geologists how to operate surveying equipment in order to make contour maps that depicted anticlinal structures. The methods of detailed structure mapping developed piecemeal over a number of years.[97] Through field work geologists hoped to identify control points by measuring the elevations of identifiable stratigraphic layers that outcropped at the surface.[98] Although the preferred instruments varied with the nature of terrain and structure in question, geologists most often adopted the telescopic alidade and plane table.[99] This device consisted of a tripod with a telescope-like instrument and a flat surface attached to the top. The geologist used the flat surface to draw the map by recording coordinates that served as the basis for contour lines that depicted the structure in two-dimensional form. To ascertain the coordinates, the geologist looked through the alidade to follow an assistant who walked over the structure with a stadia rod, which resembled a large ruler, stopping at various points for the viewer to plot measurements marked on the rod. Teaching young geology students meant showing them how to operate these instruments and to perform this procedure, so Carpenter hired Luther C. Snider, an associate from the University of Oklahoma, to train new recruits.

Learning this process required time, "at least a full year's practice," but time was not a luxury Empire could afford.[100] After experiencing significant success by mapping and drilling anticlines at Augusta and El

Dorado, the company so zealously committed its resources to expanding this approach that it acquired "an enormous amount of acreage" throughout the region, especially in Kansas and Oklahoma."[101] The pressing need to return a profit on such a large investment prevented new geologists from getting the practice necessary to learn the art of map making.

Numerous accounts confirm that Empire dramatically decreased the amount of time new geologists received in the practice of making maps and that the company's unprecedented commitment to field work failed to meet its expectations. The urgency with which Empire and its subsidiaries searched for geologists resonated in a letter that the head geologist at one of them sent to the University of Missouri's geology department. He explained that he was "much in need of some men that can do plain [sic] table with telescopic alidade." Even students who could not operate this equipment qualified for employment, as he had "been starting men who have not had previous experience at seventy five dollars per month and expenses while in the field."[102] Executives at Empire, IT10's parent company, felt the same sense of urgency to hire and train geologists, pushing them through the training process with utmost speed. One geologist related that "Diescher had a scheme that flopped of quickly training in a few weeks youth to do anticline hunting. That was about 1914."[103] The training program failed because simple anticline hunting did not always directly translate into new discoveries of oil.

Empire's massive undertaking to find surface structures started out with much promise and proved successful for a time but quickly grew into a dogmatic formula that failed to produce results. Despite considerable resources and personnel, Empire "mapped many anticlines which yielded some commercial discoveries but none of great significance."[104] As a result the "crash program" of hiring and training geologists, which lasted until 1917, "did not fulfill expectations."[105] After drilling a number of dry wells, "the large geological department then was reduced and it was realized detailed and careful field work was more important than mass production of field maps by inexperienced men, many of whom had been taken from the college class rooms before completing their regular course of study."[106] Empire had learned an important lesson that the management of other companies who mimicked their approach would not learn for a number of years, that a geological principle applied dogmatically would not substitute for "detailed and careful field work." By reducing the search for oil to mere anticline hunting, Empire overestimated the anticlinal theory's suitability for explaining all occurrences of oil and failed to consider alternative explanations.

Once oil industry geologists and executives accepted the utility of the anticlinal theory, they employed it routinely, and the creativity of many petroleum geologists diminished as they ceased to think of other geological explanations for the accumulation of oil. The anticlinal theory appealed to prospectors because it offered visual clues that guided them to potential new pools.[107] Although every hole drilled next to an anticline did not always yield oil, the faith of most geologists in the theory endured because successes at places like Cushing, Augusta, and El Dorado seemed to verify its validity.[108] Over time, however, perfunctory application of the theory retarded the rate of discovery.[109] As Wallace Pratt, one of the most prominent and successful petroleum geologists of the time, put it, "This beautiful conception, perfectly valid in principle, has often actually led us astray in the practical search for oil."[110]

Like most who had heard of the anticlinal theory, Carpenter initially conceived of it as a prospecting method with universal application and never considered how other geological phenomena might also trap oil beneath the ground. As a college student studying at the University of Oklahoma, he became "very much intrigued with Dr. I. C. White's anticlinal theory of oil and gas accumulation and bent my best efforts" to some day work as a petroleum geologist.[111] His view of this theory prompted him to enter the profession, but it changed dramatically throughout his career. In 1924 he remarked that "it never occurred to me during my university days that there was any limitation to the theory of anticlinal accumulation."[112] Thirteen years of experience in the oil industry gave him a different perspective, which he expressed in a letter to Gould, stating that "I do not regard anticlines as the only controlling factor in the accumulation of oil. There are other factors equally important."[113] He had ceased believing that the theory offered a panacea for finding oil; his was a notion some geologists would not accept until a much later date.[114] Carpenter came to realize that some geologists had ceased to think creatively and that there must be "other factors" trapping oil beneath the ground, and he determined to identify alternative explanations.[115]

By the mid-1920s Carpenter began to espouse another geological explanation for oil accumulation, the stratigraphic trap. Oil executives, however, proved reluctant to accept Carpenter's theory. While Carpenter did not dismiss the validity of the anticlinal theory, he noted that "other conditions equally important are furnished by Sand Lenses, Faults, Shoe-string Sands, etc."[116] A contemporary of Carpenter's considered "this view-point years ahead of its time." Nevertheless, management at some companies so "slavishly worships the anticlinal theory" that a geologist "has to sell an anticline to get a stratigraphic trap drilled."[117] Thus geologists misrepresented the geology of a particular site to justify the expenditure of drilling because stratigraphic traps, unlike many anticlines, offer no observable evidence at the surface.

A major advance in the discipline of petroleum geology and in the oil industry occurred when geologists articulated the principle that stratigraphy, like structures, also traps oil but far below the surface beyond human purview. By definition a stratigraphic trap consists of a discontinuity, or open space, missing from a layer of stratigraphy that provides a reservoir for oil to accumulate. The missing space results from one of several different processes.[118] For example, the reservoir may have formed from a cessation in deposition or by erosive forces that broke away a section of the stratum many years in the past.[119] Arville I. Levorsen, the petroleum geologist who most clearly articulated the concept, defined a stratigraphic trap "as one in which a variation in the stratigraphy is the chief confining element in the reservoir which traps the oil."[120] Regardless of how stratigraphic traps formed, the inability to observe or map them at the surface forced geologists to rely primarily on drillers' logs to conceptualize subsurface conditions that explained the accumulation of oil.

Long before geologists and engineers in the twentieth century began to identify limitations to the anticlinal theory, geologists on the Pennsylvania survey recognized and attempted to explain how subsurface conditions trapped oil and influenced its production. Many geologists consider John Carll, who worked under J. Peter Lesley on the Second Pennsylvania Geological Survey, a candidate for the title "father of petroleum engineering" because he compiled one of the first subsurface maps detailing the location of oil.[121] Although Carll had no university education, he drew upon data from well logs compiled by civil engineers in order to reconstruct the stratigraphy underlying Pennsylvania's Venango oil district and published these findings in 1875. Carll's report included maps that depicted cross-sections of twenty layers of stratigraphy and their position in relation to three different levels of oil reservoirs.[122] In addition to identifying the relative position of the strata, Carll included descriptions of distinctive beds and of fossils characteristic of certain depths as a guide to drillers. His practice of creating a subsurface structure map was far ahead of its time, but his greatest contribution lay in illustrating the fallacy that a relationship always exists between surface topography and the presence of oil.[123] Other geologists and engineers subsequently illustrated and articulated how subsurface conditions govern oil production, but their ideas for the most part fell upon deaf ears.[124] Many practical men and geologists so ardently desired a single comprehensible prospecting method that they held tenaciously to ideas such as the belt-line theory and later the anticlinal theory because they could employ them on the basis of surface observations alone. Once these methods failed to produce results, though, Empire Gas and Fuel Company took an unprecedented step by exploring the subsurface in a systematic and comprehensive way.

ALEX W. MCCOY

In 1917 Empire became the first company in the history of the oil industry to create a department devoted exclusively to understanding how subsurface geological conditions affected oil exploration and production. Carpenter hired Alex W. McCoy to oversee this new department. McCoy received his degree in civil engineering from the University of Missouri and in 1914 graduated with a master's degree in geology.[125] He had been teaching in the geology department at the University of Oklahoma when Carpenter hired him. The study of subsurface geology utilized all available data in order to create three-dimensional models that represented underground structures and processes and their relationship to oil.[126]

Researchers obtained data from as many sources as possible, including surface exploration, drillers' core samples, and fossils. In the 1920s geophysical exploration revolutionized the kind and quantity of data available to conduct subsurface explorations.[127] McCoy, however, started Empire's subsurface department just before that revolution took place, and during his tenure there, he initiated innovative research that would shift the industry's focus from relying solely upon geology to locate and produce oil to utilizing engineering principles as well.

Although other candidates legitimately deserve the title "father of petroleum engineering," McCoy brought recognition to this field in much the same way Carpenter had for mapping surface structures. Just as Carll's report and maps charted the subsurface of Pennsylvania's oil fields, McCoy's "group of geologists constructed subsurface structure maps, made a detailed investigation of stratigraphic problems throughout the Midcontinent district, and served generally as a research group for the Empire Companies."[128] Apparently unaware of contributions by previous geologists and engineers, McCoy never doubted that "the first attempts at petroleum engineering work, as we know it today, were made by the Empire Gas and Fuel Company in 1917," the year he began employment.[129] Although he may have overstated the case, the subsurface work he directed popularized and won acceptance for the application of engineering principles to oil exploration and production.

Initially his department focused on maximizing production at sites Carpenter had already located, as "all wells in the El Dorado and Augusta fields were drilled in, pipe and cementing recommendations submitted, and the responsibility of measuring up wells was turned over directly to this department."[130] In addition to creating and implementing more efficient procedures for extracting oil already located, McCoy's work also shifted prospectors' focus from surface to subsurface exploration.

To geologists like Carpenter, who recognized the anticlinal theory's limitations, McCoy offered promising alternatives to surface exploration by conducting research into methods that illustrated how they could find oil by shifting their focus from surface traps to stratigraphic traps. What most differentiated principles of engineering from those of geology was the less prominent role field work played in generating information. Prior to the creation of Empire's subsurface department, "the geologic evidence for guidance of operators had been predominately, if not entirely, a study of surface structure."[131] McCoy's hiring marked a turning point in this approach because it "began a period where scientific evidence used by engineers and geologists was not limited to the study of surface outcrops."[132] Instead of relying solely on field surveys, the subsurface branch of the geological department "carried on a number of experiments regarding production problems, estimates of reserves, underground stratigraphic studies, and originated many of the general practices which have been greatly improved and are now in vogue by all of the departments of geology."[133] McCoy saw his approach to exploration as a significant departure from the traditional field surveys of the kind Carpenter and other university-trained geologists conducted. He characterized his work as the starting point for a new type of exploration. McCoy differentiated "between the period when geology first became prominent throughout the Midcontinent, since it was primarily a field science of an exploratory nature, limited to the study of surface outcrops" and the period beginning in 1917 when "various phases of scientific endeavor were commenced which later have developed into the intricate phases of petroleum engineering."[134] In the first period he described, finding and producing oil had been a geological enterprise, but in the later period engineering knowledge and technical expertise determined how efficiently companies could extract oil that lay trapped beneath the ground.

In sum, the field of petroleum geology advanced rapidly throughout the first two decades of the twentieth century. As westward expansion proceeded, prospectors guided by surface indications such as seeps and anticlines rapidly located and depleted shallow reserves of oil. Both practical men and petroleum geologists participated in the location and extraction of shallow oil deposits, but the authority and power of each type of practitioner began to change as the need to explore deeper beneath the surface increased. Doherty recognized the need to invest in industrial research and development in order to innovate in subsurface exploration and production as well as the limited ability of craft practitioners to cultivate increasingly sophisticated but necessary science and technology. Geological principles taught at universities and training in field experience provided students with an epistemology that they could build upon and elaborate in order to interpret subsurface geology, offer new explanations of oil accumulation, and answer the call to revolutionize oil prospecting. Doherty included the expertise of this emerging middle-class elite in his technological system to facilitate innovation, setting a new standard for the role of science and technology within the oil industry that other companies eventually adopted. The integration and consolidation of the oil industry by these companies greatly affected smaller producers, who lacked the capital to operate on large economies of scale. Smaller producers, who struggled to compete against integrated companies, interpreted the industry's consolidation as a monopolistic threat and turned to the state for a remedy to their plight.

Conclusion

At 4:30 a.m. on October 29, 1917, a nitroglycerine bomb exploded in an upscale neighborhood of Tulsa, Oklahoma, where J. Edgar Pew, president of Carter Oil Company, and his family slept in their home. The International Workers of the World (IWW) planted the bomb to murder Pew and his family as the first in a series of planned attacks to kill executives of major oil companies in the area.[1] The Tulsa newspaper reported that the IWW "believes that by blowing up homes of prominent oil men, dynamiting oil tanks, refineries, and derricks and generally creating a reign of terror they will sufficiently frighten the employers to accomplish their purposes."[2] Pew and his family escaped without harm, but the Wobblies hoped that acts of domestic terrorism would convey to other executives of large integrated oil companies that workers would not passively allow organizational changes within the industry to strip them of their power to profit as laborers. The Carter Oil Company was one of many subsidiaries resulting from the Supreme Court ruling that broke up Standard Oil Company, and people in western states whose livelihoods depended on oil remained on guard to prevent Standard from reasserting its power in new guises. At the time of the bombing, the Wobblies called for their members to descend upon the Oklahoma oil fields, which served as "the center of wealth controlled by the Standard Oil company, where work is plentiful and labor scarce."[3] The Wobblies knew what Standard's architect

John D. Rockefeller had already learned, that whoever controlled oil possessed a great deal of power. This lesson presented itself to new generations as the oil industry moved westward.

The professional expertise of geologists and engineers translated into significant control for system-builders like Henry Doherty who constructed large integrated companies that disempowered many small-scale oil men who failed to adapt to the organizational changes occurring within an emergent economy of scale. Opportunities within the oil industry changed dramatically as the fields of petroleum geology and engineering found their way into university curricula and corporate research and development departments. Some practical men like Tom Slick accommodated the new science and technology, but others either could not afford such capital-intensive business practices or remained wedded to methods of locating and producing oil that relied solely on physical labor and tacit knowledge that they failed to reproduce in translocal environments. As the industry moved westward throughout the first half of the twentieth century into western states like Kansas, Oklahoma, Texas, and California, geologists and engineers learned how to replicate, innovate, and systematically apply geological theories that enhanced their authority and gradually began to disempower practical men.

Given the central importance of oil in the United States and global economies, it hardly seems novel that people working in the oil industry have involved themselves in struggles for power to control the knowledge governing this increasingly valuable and scarce natural resource. Throughout the industrialized world, people bear witness daily to the dependency of their lives on oil and to an industry that produces a resource they consume in abundance. Oil's central importance and escalating cost lead many consumers to see the industry as an all-powerful and monolithic entity that exerts significant influence over their lives, economically, politically, and diplomatically as well as in myriad other ways. Certainly, the largest oil companies and the executives these companies employ possess and wield a great deal of power that is sometimes exercised very injudiciously, but historical depictions of an industry that acted autonomously and with a single-minded pursuit of economic power only oversimplify the diversity of choices made by people who worked to locate and produce oil. What was at stake in the battle for power by a single interest group remained situated in the exigencies of time and place. This study has shown that oil prospectors as a category of laborers in the industry did not always see their work in nature in the same way and that their perspectives sometimes clashed, but their epistemological differences often drove innovation in petroleum geology and engineering.

Prospectors fought one another intensely because they defined nature differently and because they understood that significant power befell whoever's definition prevailed. The issue of who gets to define nature has significant consequences for the lives of people living and working closest to the resources they produce and consume.[4] To understand how and why the definitions of prospectors differed, we must understand that no clear distinctions existed between the worlds of nature and artifice.[5] Oil prospectors did not differentiate clearly or consistently between the natural world, which housed the oil they sought, and the world of scientific and technological arti-facts, which facilitated locating and extracting oil.[6] Rather, prospectors functioned both as producers and consumers of the landscapes they traversed, studied, and struggled to understand. They produced oil at the same time they consumed landscapes aesthetically, intellectu-ally, and culturally. Field work offered opportunities for both work and play, and each prospector marked the boundaries between these ways of encountering nature differently. Their interactions with diverse terrain and geology imparted a sense of place at the same time these interactions facilitated the practical tasks that led prospectors to oil and potential economic gain. Prospectors encountered nature with a variety of motivations, aspirations, and desired outcomes. Sometimes they wanted profit, other times to conquer Mother Nature or to verify geological theories by observing them firsthand. Given their range of motivations, prospectors' methods overlapped and complicated their relationships to nature in often paradoxical ways, blurring their activities as producers and consumers of nature.

Because prospectors understood their work differently, I have often wondered what many of the central characters in this history may have thought about witnessing a gusher firsthand and about the gradual demise of the phenomenon as the oil industry matured. J. Peter Lesley may have had no familiarity with a gusher whatsoever because of the time and place that he practiced petroleum geology. Pennsylvania's geology produced oil that seeped rather than gushed to the surface. Geological structures known as stratigraphic traps captured oil without the natural gas that accumulated in anticlines farther west and that expelled oil in the form of geysers. Although Tom Slick was also from Pennsylvania, he would have recognized a gusher because he followed the industry westward to new environments on the southern plains, where he could learn, or "sort of sense, by intuition, where there ought to be oil."[7] Environment influenced how he cultivated knowledge just as it influenced Lesley, but the best practical men and geologists adeptly adjusted to new geological contexts. Charles Gould might have reacted to a gusher with the same ambivalence he approached the oil industry generally. On the one hand, he would have admired the grandeur of gushers and the geological forces that created them, but on the other, his proclivity for rationalizing resources along scientific principles would have prompted him to lament the economic waste of a remunerative natural resource. In the end he might have both written a poem about gushers and used them to illustrate geological principles to his students, who could parlay their knowledge into jobs as petroleum geologists. If Gould might have demonstrated ambivalence toward gushers, Henry L. Doherty almost certainly would have responded with the most dread and regret but also an organized, efficient, and well-engineered measure to contain the loss of oil. Doherty understood oil as a natural resource that made up part of a larger cultural and natural landscape that included people who worked in science, industry, and government and all of whose activities had to be organized to maximize efficiency. Whoever the practitioner, his conception of gushers would have reflected his understanding of how to acquire authority and power over oil and nature's geological properties.

Gushers did not disappear entirely from the American landscape in 1920, but their days were numbered as geologists and engineers elaborated upon the science and technology used to find and extract oil. In the 1920s, geologists and physicists began to innovate in finding oil with seismic waves. They produced increasingly sophisticated and arcane geological knowledge and revolutionized how the industry found oil. Still, practical men continued to play an important role in prospecting into the 1930s, locating enormous pools of oil in Oklahoma, Texas, and California. Many of these discoveries erupted as gushers, but the volume of oil they released onto the market began presenting more problems and instability than excitement and romance. Overproduction destabilized the industry so seriously throughout the twenties and thirties that oil producers proposed, debated, and contested various attempts to regulate their industry at the state and federal levels in order to "conserve" oil. Struggles for control and power within the oil industry grew even more intense as companies produced more oil than markets could consume and the price plummeted. Petroleum geology and engineering figured into some of the conservation plans proposed, albeit to varying extents. Scientific and technological advances aided in containing and controlling gushers, but the question of how best to conserve oil constituted another struggle for power in which a diversity of voices would clash.

Notes

INTRODUCTION

1. Yergin, *The Prize*, 12.

2. For the best overview of the literature and related themes, see Richard White, "From Wilderness to Hybrid Landscapes." Other works in this vein include Fiege, *Irrigated Eden*; Langston, *Where Land and Water Meet*; Morse, *Nature of Gold*; Jennifer Price, *Flight Maps*; Sutter, *Driven Wild*.

3. Richard White, "From Wilderness to Hybrid Landscapes."

4. Morse, *Nature of Gold*. For a similar analysis of the oil industry as both a natural and cultural creation, see Black, *Petrolia*, 61.

5. Richard White, "From Wilderness to Hybrid Landscapes," 560–61.

6. Richard White, "'Are You an Environmentalist?'" Many environmental historians demonstrate nature's agency in shaping history; see the following for some of the best examples: Pyne, *Fire in America*, *The Ice*, and *How the Canyon Became Grand*.

7. Richard White, "'Are You an Environmentalist?'" For an excellent overview of the factors mediating work and the environment, see Andrews, *Killing for Coal*, 125, 331n5; McEvoy, "Working Environments."

8. Andrews, *Killing for Coal*, 125.

9. I purposely make no clear distinction between "objective" and "subjective" knowledge and adopt the approach of other scholars who treat "objectivity" as a cultural construct professionals used to legitimize their practices, achieve consensus, and often to delegitimize competing paradigms. See Linda Nash, "Changing Experience of Nature." Although the historical actors who appear in this story used terms like "objective," "subjective," "practical," and "theoretical" to describe the kinds of information they generated, I assume that all prospecting knowledge arose from physical, sensory, and intellectual encounters with nature. Indeed, no single constituency or type of knowledge held the key to finding oil. See Linda Nash, "Changing Experience of Nature," 1602.

10. Shapin, *Scientific Life*.

11. Shapin, *Scientific Life*, 9.

12. Linda Nash, "Changing Experience of Nature," 1602.

13. Bowker and Star, *Sorting Things Out*, 10.

14. Bowker and Star, *Sorting Things Out*, 5.

15. Bowker and Star, *Sorting Things Out*, 10.

16. Orr, *Talking about Machines*.

17. Cognitive anthropologists have theorized much about the "communities of practice" in which people learn through collaboration. See Lave and Wenger, *Situated Learning*; McDermott, Snyder, and Wenger, *Cultivating Communities of Practice*.

18. Lucier, *Scientists and Swindlers*.

19. The seminal transitional figure is Elmer Sperry, who concentrated his work on the electrical industry but worked with mining machinery, automobiles, streetcars, and other technology. Doherty too began his inventing career in the electric industry, but I am more concerned with how he applied ideas he learned in that field to the oil industry. See Hughes, *Elmer Sperry*, xiv.

20. The literature on professionalization is vast. For the best overview of the historiographic trends I cite here, see Brown, "Profession." See also Larson, *Rise of Professionalism*; Ben-David, *Scientist's Role in Society*; Bledstein, *Culture of Professionalism*; Brown, *Definition of a Profession*; Daniels, "Professionalization in American Science"; Geison, *Professions and Professional Ideologies*; Haskell, *Authority of Experts*; Hobson, "Professionals, Progressives, and Bureaucratization"; Terence Johnson, *Professions and Power*; Kimball, *"True Professional Ideal" in America*; Veysey, "Who's a Professional?"

21. Abbott, *System of Professions*, 2.

22. Pamela Smith, "Science on the Move," 346.

23. Pamela Smith, "Science on the Move," 348.

24. Shapin, "Placing the View." Shapin contends that "we need to understand not only how knowledge is made in specific places but also how transactions occur between places" (6–7).

25. Arthur Johnson, *Petroleum Pipelines and Public Policy*, 19. Gerald White also notes that the "substantial geographic isolation of the California industry from the rest of the nation" has shaped the historiography on the topic ("California's Other Mineral," 136n2).

26. The fact that much of California's oil sat beneath public lands also differentiates that state's history from states like Pennsylvania, Kansas, Oklahoma, and Texas, where oil production took place primarily on private lands. See Sabin, *Crude Politics*, 10. More history remains to be written on the development of petroleum geology in California, but a good starting point should include the following:

Owen, *Trek of the Oil Finders*; Gerald White, "California's Other Mineral" and *Formative Years in the Far West*; Lucier, *Scientists and Swindlers*.

27. Pamela Smith, "Science on the Move," 348.

28. Literature on the social construction of technology, or SCOT, is vast. For a good starting point, see Bijker, Hughes, and Pinch, *Social Construction of Technological Systems*. The intersection of environmental history and the history of technology comprises a specific area of inquiry within SCOT. For a good overview, see Stine and Tarr, "At the Intersection of Histories."

29. Linda Nash, "Changing Experience of Nature," 1601–3; Pamela Smith, "Science on the Move," 346.

30. Hounshell, "Hughesian History of Technology," 214.

31. Actor-network theory originated with the work of Bruno Latour and Michel Callon but has been elaborated by numerous other scholars. For an introduction to the theory and overview of subsequent scholarship, see Czarniawska-Joerges and Hernes, *Actor-Network Theory and Organizing*; Latour, *Reassembling the Social*; Law and Hassard, *Actor Network Theory and After*.

32. Richard White, *Organic Machine*, ix–x.

1. VERNACULAR AUTHORITY IN THE OIL FIELD

1. Union Oil Company of California, *Fortune*, May 1940, 4.

2. I borrow the term "vernacular" from Katherine Pandora, who used it to make the point that "there exists a range of activities, forms of thought, modes of expression that can be understood as 'vernacular natural knowledge,' and that this knowledge is inevitably entangled with the elite science that is the conventional object of our scrutiny" ("Natural Histories of Science"). For additional examples of how vernacular knowledge influences the study of science, see Pandora, "Knowledge Held in Common."

3. There were three major categories of consultants employed by the industry — scientists, engineers, and practical oil men — but there was no single type for each category. "In sum, a socially and intellectually diverse collection of experts flourished during the early years of the petroleum industry" (Lucier, "Scientists and Swindlers," 390–91). See also Lucier, *Scientists and Swindlers*.

4. Historians have written much about industry's integration of scientific experts. The chemical and electric industries established the pattern others would follow. See Wengenroth, "Science, Technology, and Industry," 252.

5. Cahan, "Looking at Nineteenth-Century Science," 3–8; Pauly, "Science," 615.

6. Paul Lucier argues in his various works that scientists working as consultants figured prominently in the coal and oil businesses throughout the nineteenth century and through their efforts blurred the lines between science, industry, and

society. See "Scientists and Swindlers," *Scientists and Swindlers*, and "Commercial Interests and Scientific Disinterestedness."

7. There is an enormous literature on craft knowledge throughout the nineteenth and early twentieth centuries. For a good introduction to this work, see Carlson, *Innovation as a Social Process*, 4–5. For the interrelationships between science and industry, see Lucier's work and Wengenroth, "Science, Technology, and Industry."

8. Gorman, "Types of Knowledge," 219.

9. Historians of science and technology have produced a great body of work on tacit knowledge. See Polanyi, *Tacit Dimension*; Collins, "TEA Set"; Sorensen and Levold, "Tacit Networks."

10. "Local knowledge" is one of those phrases whose meaning seems self-apparent, but its simple appearance veils a more complex and deeper significance. Clifford Geertz offers the best discussion of the term but does not provide a clear-cut definition. The closest he comes to defining the term occurs with his discussion of law as local knowledge, by which he means "vernacular characterizations of what happens connected to vernacular imaginings of what can" (*Local Knowledge*, 215). James Scott sees local knowledge on a par with practical knowledge and know-how as indispensable ways of knowing the world that stand in contrast to knowledge that is more formal and deductive. See Scott, *Seeing Like a State*, 6, 24.

11. Oldroyd, "Earth Sciences," 106–7. Oldroyd notes that most studies of the relationship between laymen, or "the geological underworld," and so-called elite practitioners of geology focus mostly upon the discipline during the British industrial revolution. Most studies of British geology emphasize the earlier nineteenth century, and studies of the latter half of the century have barely begun.

12. The theme of rationality in contradistinction to good fortune and luck runs throughout the work of cultural historian Jackson Lears. See Lears, *Something for Nothing*, 130–31.

13. Valencius, *Health of the Country*, 159–61. For an in-depth discussion of the relationship between local knowledge and anthropology, see Geertz, *Local Knowledge*. Local knowledge allows people to improvise when unpredictable circumstances occur, an option greatly restricted by more formal and less flexible deductive reasoning. As a result local knowledge undermined state officials' high-modernism that aimed to objectify, codify, and administer nature (Scott, *Seeing Like a State*, 6, 24).

14. Landes, *Petroleum Geology*, 21–29.

15. Locating oil by sense of smell was a valuable technique, but the idea of oil smelling became a generic term employed for a variety of searching methods (Boatright, *Folklore of the Oil Industry*, 16).

16. Landes, *Petroleum Geology*, 21.

17. Owen, *Trek of the Oil Finders*, 6. For a list and description of the five basic

types of surface indications, see Levorsen, *Geology of Petroleum*, 15; Clapp, "Occurrence of Petroleum."

18. Sarewitz, Pielke, and Byerly, "Introduction," 2.

19. Oreskes, "Why Predict?," 23.

20. Oreskes, "Why Predict?," 23–24.

21. Sarewitz, Pielke, and Byerly, "Introduction," 2.

22. Boatright, *Folklore of the Oil Industry*, 66

23. For the high probability of finding oil in the early days of an unexploited oil region, see Boatright and Owens, *Tales from the Derrick Floor*, 96. Oil was so abundant in the early days of the Appalachian oil fields that "hit-and-miss methods found enough of them to glut the market" (Ball, Ball, and Turner, *This Fascinating Oil Business*, 46–47; Boatright, *Folklore of the Oil Industry*, 20).

24. Lears, *Something for Nothing*.

25. Lears, *Something for Nothing* 3.

26. Lears, *Something for Nothing* 231.

27. Quoted in Miles, *King of the Wildcatters*, 68. When DeGolyer asked his friend and leading petroleum geologist of the early twentieth century Wallace Pratt what accounted for his success, Pratt answered, "If I were to reply that I had simply been lucky you would charge me with undue modesty; but my 'successes' do appear to me to be largely fortuitous" (Wallace E. Pratt to Everette DeGolyer, February 23, 1945, box 12, file 1513, Everette DeGolyer Collection, Southern Methodist University [hereafter cited as EDC]).

28. Fabian, *Card Sharps*.

29. Lears, *Something for Nothing*, 19.

30. Lears, *Something for Nothing*, 19; Spence, *Mining Engineers and the American West*, 110–14; Young, *Western Mining*, 40–52.

31. Quoted in Boatright, *Folklore of the Oil Industry*, 93.

32. Olien and Olien, *Easy Money*; Boatright, *Folklore of the Oil Industry*, 94.

33. Blakey, *Oil on Their Shoes*, 18–22; Copithorne, "From Doodlebug to Seismography," 44.

34. Weber, *Economy and Society*, 2:956–58, 1112.

35. Weber, *Economy and Society*, 2:1111–13.

36. Lucier, "Scientists and Swindlers," 418, and *Scientists and Swindlers*.

37. Wright, *Oil Regions of Pennsylvania*, 61.

38. Bone, *Petroleum and Petroleum Wells*, 35.

39. Wright, *Oil Regions of Pennsylvania*, 63.

40. Wright, *Oil Regions of Pennsylvania*, 62–63; Lucier, "Scientists and Swindlers," 418–19.

41. *New York Times*, October 8, 1922.

42. Quoted in Boatright, *Folklore of the Oil Industry*, 21.

43. Boatright, *Folklore of the Oil Industry*, 24.

44. Ruth Bryan, interview by Robert A. Montgomery, August 2, 1959, interview no. 207, box 3K23, file 60, Oral History of the Oil Industry Collection, Center for American History, Austin TX (hereafter cited as OHC).

45. Bryan interview.

46. Guy Finley, interview by W. A. Owens, May 5, 1956, interview no. 182, box 3K23, files 1–34, OHC.

47. Landes, *Petroleum Geology*, 8.

48. Boatright and Owens, *Tales from the Derrick Floor*, 14, 21; Santschi, *Modern "Divining Rods,"* 53.

49. Tait, *Wildcatters*, 73.

50. Santschi, *Modern "Divining Rods,"* 53; Meinzer, Introductory Note, 5–6.

51. Silliman, "Divining Rod," 212.

52. Geertz, *Local Knowledge*, 76.

53. Geertz, *Local Knowledge*, 78.

54. Geertz, *Local Knowledge*, 78.

55. Landes, *Petroleum Geology*, 8.

56. Boatright, *Folklore of the Oil Industry*, 43–44.

57. Alexander Balfour Patterson, oral interview excerpted in Boatright and Owens, *Tales from the Derrick Floor*, 17–18. The entire interview is located in box 3K22, files 126–27, OHC. For more on Dr. Griffith, see also in this collection the oral history of Alan W. Hamill, box 3K21, files 84–85.

58. Patterson, in Boatright and Owens, *Tales from the Derrick Floor*, 18.

59. Santschi, *Modern "Divining Rods,"* 61.

60. O. W. Killam, box 3K23, file 183, OHC, and in Boatright and Owens, *Tales from the Derrick Floor*, 20.

61. Killam, OHC, and Boatright and Owens, *Tales from the Derrick Floor*, 20.

62. Polanyi, *Tacit Dimension*, 14–15.

63. Polanyi, *Tacit Dimension*, 6.

64. Polanyi, *Tacit Dimension*, 6.

65. Polanyi, *Tacit Dimension*, 12.

66. Polanyi, *Tacit Dimension*, 13.

67. Polanyi, *Tacit Dimension*, 14–15.

68. Killam, OHC.

69. Geertz, *Local Knowledge*, 76.

70. Killam, OHC.

71. Killam, OHC.

72. Valencius, *Health of the Country*, 53.

73. Barton, "Wigglestick," 312.

74. Barton, "Wigglestick," 312.

75. Barton, "Wigglestick," 312.

76. Barton, "Wigglestick," 312.

77. Lears, *Something for Nothing*, 11–12.

78. Lucier, *Scientists and Swindlers*, 252–54.

79. Turner, *Significance of the Frontier*, 37.

80. Turner, *Significance of the Frontier*, 297.

81. Turner, *Significance of the Frontier*, 318.

82. Turner, *Significance of the Frontier*, 318.

83. Wright, *Oil Regions of Pennsylvania*, 58.

84. Mellon, *Judge Mellon's Sons*, 158.

85. Mellon, *Judge Mellon's Sons*, 158.

86. Mellon, *Judge Mellon's Sons*, 158. Boatright and Owens support the view that wildcatters relied upon an eclectic mix of oil-finding methods: "The wildcatter was likely to rely on a strange mixture of surface observation, superstition, the occult, and even the divine" (*Tales from the Derrick Floor*, 14).

87. Mellon, *Judge Mellon's Sons*, 158.

88. Mellon, *Judge Mellon's Sons*, 158.

89. Mellon, *Judge Mellon's Sons*, 158.

90. Mellon, *Judge Mellon's Sons*, 158.

91. Boatright, *Folklore of the Oil Industry*, 5–6.

92. Mellon, *Judge Mellon's Sons*, 159.

93. Boatright, *Folklore of the Oil Industry*, 9.

94. Boatright, *Folklore of the Oil Industry*, 9. Boatright cites examples of the folk traditions that led congregations to lease out their churchyard cemeteries.

95. Blakey, *Oil on Their Shoes*, 18.

96. Edward Bloesch, "Early Day Petroleum Geology in Oklahoma," box 3A, file "Ed Bloesch Letters," Edgar Wesley Owen Collection, American Heritage Center, University of Wyoming (hereafter cited as EOC).

97. I use the date 1885 because that is the year that a geologist first demonstrated geology's practical application in finding oil and gas and published his findings. This accomplishment marked a pivotal point that will be elaborated upon in subsequent chapters. See Israel White, "Geology of Natural Gas" and "Mannington Oil Field."

98. Tait, *Wildcatters*, 75.

99. For a more extensive discussion of reservoir rocks, see "Reservoir Rock," Memoir 5, in Dott and Reynolds, *Sourcebook for Petroleum Geology*.

100. DeGolyer, "Concepts on Occurrence of Oil and Gas," 24; Tait, *Wildcatters*, 75.

101. Carll, "Geology of the Oil Regions," 245.

102. DeGolyer, "Concepts on Occurrence of Oil and Gas," 24.

103. DeGolyer, "Concepts on Occurrence of Oil and Gas," 22; Owen, *Trek of the Oil Finders*, 66.

104. Fuller, "Appalachian Oil Field," 625–26.

105. C. E. Bishop, quoted in Henry, *Early and Later History of Petroleum*, 486–92.

106. Frederick Prentice, who was briefly associated with Angell, first observed the belt pattern in 1861 and deserves partial credit for the theory. See Owen, *Trek of the Oil Finders*, 102, 104; Dott and Reynolds, *Sourcebook for Petroleum Geology*, 411.

107. Owen, *Trek of the Oil Finders*, 105.

108. Bishop, "Vacation in Petrolia," 670.

109. Bishop, "Vacation in Petrolia," 670.

110. Bishop, "Vacation in Petrolia," 670.

111. Bishop, "Vacation in Petrolia," 670.

112. Owen, *Trek of the Oil Finders*, 102; DeGolyer, "Concepts on Occurrence of Oil and Gas," 23; Fuller, "Appalachian Oil Field," 625.

113. Even one of the largest oil companies in Pennsylvania explored for oil by following lines without a geological justification. See Fuller, "Appalachian Oil Field," 626.

114. Owen, *Trek of the Oil Finders*, 102.

115. Geertz, *Local Knowledge*, 75.

116. Geertz, *Local Knowledge*, 76.

117. Geertz, *Local Knowledge*, 76.

118. Geertz, *Local Knowledge*, 79.

119. Geertz, *Local Knowledge*, 78.

120. Geertz, *Local Knowledge*, 80.

121. Meinzer, Introductory Note, 5.

122. Sewell, "Concept(s) of Culture," 39.

2. COLLABORATIVE AUTHORITY

1. Oleson and Voss, Introduction; Kohlstedt, "Geologists' Model for National Science." For discussions of the gentleman scientist as a stock figure in American history, see Daniels, *Science in American Society*, 45–46, and *American Science in the Age of Jackson*, 63–85; Cannon, *Science in Culture*, 73. Gentleman scientists and other lay practitioners were often dubbed amateur scientists by their contemporaries. I refrain from using this term to avoid suggesting that their knowledge was qualitatively inferior. For additional discussions of so-called amateur knowledge, see the following: Bruce, *Launching of Modern American Science*, 35–36, 135–36; Daniels, *American Science in the Age of Jackson*, 19–20, as well as chaps. 2–4; Rudwick, *Great Devonian Controversy*, 17–18.

2. For a similar description of agricultural science, see Cohen, *Notes from the Ground*. Many historians of engineering have documented the role that informal, tacit, and rule-of-thumb knowledge played in the formation of science and engineering in particular. See Calvert, "Search for Engineering Unity," 48–49; Spence, *Mining Engineers and the American West*, 20–21, 34–35.

3. Cohen, *Notes from the Ground*, 8.

4. Cohen, *Notes from the Ground*, 84.

5. Cohen, *Notes from the Ground*, 8.

6. Rudwick, *Great Devonian Controversy*, 6.

7. Wrigley, "Present and Future of the Pennsylvania Oil Fields," 186.

8. Wrigley, "Present and Future of the Pennsylvania Oil Fields," 186.

9. Hineline, "Visual Culture of the Earth Sciences," 1–28; Nystrom, "Learning to See."

10. Rudwick, "Emergence of a Visual Language," 149–50, 177. See also Carol Johnson, "Evolution of Illustrated Texts."

11. Rudwick, "Emergence of a Visual Language," 177–78.

12. Aldrich, "American State Geological Surveys," 133.

13. Hendrickson, "Nineteenth-Century State Geological Surveys," 363. For more on the politics involved in funding a survey, see Laudan, *From Mineralogy to Geology*, 295.

14. Hendrickson, "Nineteenth-Century State Geological Surveys," 363.

15. Hendrickson, "Nineteenth-Century State Geological Surveys," 366–67. Geologists at times displayed the view that they possessed a broader, less myopic vision than legislators in determining how to organize the surveys. For an example, see "State Geological Surveys and Economic Geology," 377.

16. Despite his bad luck with Iowa and Wisconsin, Hall had better luck with other attempts to acquire appropriations for purely scientific work. See "State Geological Surveys and Economic Geology," 367, 376.

17. Hendrickson, "Nineteenth-Century State Geological Surveys," 367.

18. Hendrickson, "Nineteenth-Century State Geological Surveys," 368.

19. Newell, "American Geologists and Their Geology," xv.

20. Adams, *Old Dominion, Industrial Commonwealth*, 122.

21. Owen, *Trek of the Oil Finders*, 46.

22. Henry D. Rogers to William B. Rogers, May 1, 1841, in Emma Rogers and W. T. Sedgwick Rogers, *Life and Letters of William B. Rogers*, 1:190. See also Hendrickson, "Nineteenth-Century State Geological Surveys," 370; Boscoe, "Insanities of an Exalted Imagination."

23. Henry Rogers, *First Annual Report of the State Geologist*, 22; quoted in Millbrooke, "Henry Darwin Rogers," 72.

24. Henry Rogers, *First Annual Report of the State Geologist*, 22.

25. Henry Rogers, *First Annual Report of the State Geologist*, 22; Hendrickson, "Nineteenth-Century State Geological Surveys," 365–66.

26. Lesley, "History of the First Geological Survey," 70.

27. Lesley, "History of the First Geological Survey," 71.

28. Lesley, "History of the First Geological Survey," 71.

29. Lesley, "History of the First Geological Survey," 72–73.

30. Lesley, "History of the First Geological Survey," 72.

31. Lesley, "History of the First Geological Survey," 72.

32. Lesley, "History of the First Geological Survey," 111.

33. Lesley, "History of the First Geological Survey," 111.

34. Lesley, "History of the First Geological Survey," 111.

35. Lesley, "History of the First Geological Survey," 111.

36. Lesley, "History of the First Geological Survey," 111.

37. Lesley, "History of the First Geological Survey," 112.

38. Lesley, "History of the First Geological Survey," 111–12; Adams, *Old Dominion, Industrial Commonwealth*, 120, 137–38; Gerstner, *Henry Darwin Rogers*.

39. Adams, "Partners in Geology," 18.

40. Adams, "Partners in Geology," 15.

41. Adams, "Partners in Geology," 18.

42. Lesley, *Manual of Coal*, 23.

43. Lesley, *Manual of Coal*, 23.

44. Lesley, *Manual of Coal*, 124.

45. Lesley, "History of the First Geological Survey," 79.

46. Ancient Babylonians and Greeks practiced topography, as did Europeans during the Renaissance. Topographical mapping entered a modern phase beginning in approximately 1700 and lasting to 1950 during which period optical instruments that contained lenses and mirrors allowed their users to measure distances as well as vertical and horizontal angles. See Hatzopoulos, *Topographic Mapping*, 4–6.

47. Lesley, "History of the First Geological Survey," 79.

48. Lesley, "History of the First Geological Survey," 79.

49. Lesley, "History of the First Geological Survey," 79.

50. Lesley, *Manual of Coal*, 123.

51. Lesley, *Manual of Coal*, 187.

52. Lesley, *Manual of Coal*, 188.

53. Lesley, *Manual of Coal*, 188.

54. Lesley, *Manual of Coal*, 191.

55. Lesley, *Manual of Coal*, 191.

56. Lesley, *Manual of Coal*, 192.

57. Lesley, *Manual of Coal*, 192.

58. Lesley, *Iron Manufacturer's Guide*, ix.

59. Lesley, *Iron Manufacturer's Guide*, x.

60. Gerstner, *Henry Darwin Rogers*, 98.

61. William Rogers, *A Few Facts Regarding the Geological Survey of Pennsylvania*, 4.

62. William Rogers, *A Few Facts Regarding the Geological Survey of Pennsylvania*, 4.

63. William Rogers, *A Few Facts Regarding the Geological Survey of Pennsylvania*, 4.

64. Lucier, *Scientists and Swindlers*, 125.

65. William Rogers, *A Few Facts Regarding the Geological Survey of Pennsylvania*, 13.

66. Lesley, *Iron Manufacturer's Guide*, xi.

67. William Rogers, *A Few Facts Regarding the Geological Survey of Pennsylvania*, 13.

68. Rudwick, "Emergence of a Visual Language," 164.

69. William Rogers, *A Few Facts Regarding the Geological Survey of Pennsylvania*, 13.

70. Rudwick, "Emergence of a Visual Language," 177–78.

71. Lesley, *Iron Manufacturer's Guide*, xi.

72. Lesley, *Iron Manufacturer's Guide*, xi.

73. Lesley, "Second Geological Survey [1874]," ix.

74. Lesley, "Second Geological Survey [1874]," ix.

75. Lesley, "Second Geological Survey [1874]," ix–x.

76. J. Peter Lesley, "Letter from Professor Leslie; Reprinted in *Monogahela Republican*," August 21, item 19, diary vol. 4, 1877–1881, J. P. Lesley Collection, American Philosophical Society, Philadelphia (hereafter cited as JLC).

77. "Editorial, *Monogahela Republican*," August 21, item 19, diary vol. 4, 1877–1881, JLC.

78. Lesley, "Letter from Professor Leslie."

79. *Dictionary of American Biography*, s.v., "Lesley, Peter."

80. *Dictionary of American Biography*, s.v., "Lesley, Peter."

81. J. Peter Lesley to Benjamin S. Lyman, September 25, 1865, ser. I, correspondence, JLC.

82. Lesley to Lyman, September 25, 1865.

83. Lesley to Lyman, September 25, 1865.

84. Lesley to Lyman, September 25, 1865.

85. Lesley, "Second Geological Survey [1875]," xxii.

86. Lesley, "Second Geological Survey [1875]," xxii.

87. Lesley, "Second Geological Survey [1875]," xx.

88. Curiously, William Morris Davis makes this point but simultaneously argues that the blame should lay with the principle itself rather than with Lesley. See Davis, "Biographical Memoir of J. Peter Lesley," 217.

89. Lesley, "Second Geological Survey [1875]," xx.

90. Stevenson, "J. Peter Lesley," 2–3.

91. Stevenson, "Memoir of J. Peter Lesley," 535.

92. Stevenson, "J. Peter Lesley," 2–3.

93. Benjamin Smith Lyman, quoted in Ames, *Life and Letters of Peter and Susan Lesley*, 2:131.

94. Lyman, quoted in Ames, *Life and Letters of Peter and Susan Lesley*, 2:131.

95. Chance, "Biographical Notice," v.

96. Chance, "Biographical Notice," v.

97. Chance, "Biographical Notice," v.

98. Pees, "Early Oil and Gas Exploration," 111.

99. J. P. Lesley, August 16, 1879, item 19, diary vol. 4, 1877–1881, JLC.

100. Lesley, August 16, 1879.

101. Lesley, August 16, 1879.

102. J. Peter Lesley to Israel Charles White, November 23, 1880, JLC.

103. Lesley to White, November 23, 1880.

104. Lesley to White, November 23, 1880.

105. Lesley to White, November 23, 1880.

106. Lesley to White, November 23, 1880.

107. Lesley to White, November 23, 1880.

108. Lesley to White, November 23, 1880.

109. Lesley to White, November 23, 1880.

110. Chance, *Northern Townships of Butler County*, 155.

111. Chance, *Northern Townships of Butler County*, 155.

112. Chance, *Northern Townships of Butler County*, 154–55.

113. Chance, *Northern Townships of Butler County*, 155.

114. Chance, *Northern Townships of Butler County*, 155.

115. Wrigley worked as an engineer for Ulysses S. Grant during the Civil War, published in the *Engineering and Mining Journal*, and designed many houses for the oil industry elite in Pennsylvania. See Sherretts and Moore, *Oil Boom Architecture*, 8; Simon, *Papers of Ulysses S. Grant*, 27:146.

116. Wrigley, "Present and Future of the Pennsylvania Oil Fields," 186.

117. Wrigley, "Present and Future of the Pennsylvania Oil Fields," 186.

118. Wrigley, Jones, and Lesley, *Special Report on the Petroleum of Pennsylvania*, 7.

119. Israel White, "Geology of Natural Gas."

120. Other geologists had articulated the anticlinal theory before White, but none had demonstrated its practical application. They are referenced in my subsequent discussion.

121. Most introductory-level geology books provide explanations of anticlines. See Ball, Ball, and Turner, *This Fascinating Oil Business*, 49; Anderson, *Fundamentals of the Petroleum Industry*, 89.

122. Ball, Ball, and Turner, *This Fascinating Oil Business*, 55; Levorsen, "Stratigraphic Versus Structural Accumulation," 524.

123. Dickey, "Oil Is Found with Ideas" and "Pennsylvania's Contribution to Petroleum Geology," 1143; Harper, "Incredible John F. Carll."

124. *Derrick's Hand-Book of Petroleum*, 2:520.

125. Carll, *Report of Progress*, 23.

126. Carll, *Geology of the Oil Regions*, 175.

127. Carll, *Geology of the Oil Regions*, 175.

128. Carll, *Geology of the Oil Regions*, 176.

129. Carll, *Report of Progress*, 23.

130. Carll, *Report of Progress*, 23.

131. Harper, "Incredible John F. Carll," 4.

132. Carll, *Report of Progress*, 16.

3. SHARED AUTHORITY

1. Roswell H. Johnson to James A. Veasey, February 24, 1941, "Belt-line Theory," box 23, James A. Veasey Collection, American Heritage Center, University of Wyoming (hereafter cited as JVC).

2. Johnson to Veasey, February 24, 1941.

3. Johnson to Veasey, February 24, 1941.

4. One of the most reliable sources for the history of Oklahoma petroleum geology dates the year of acceptance at 1913. See Powers, "Petroleum Geology in Oklahoma," 5. DeGolyer also sees 1913 as a pivotal year for dating industry's acceptance of geology. See DeGolyer, *Development of the Art of Prospecting*, 28.

5. Wiebe, *Search for Order*.

6. Scott, *Seeing Like a State*, 326–27.

7. Scott, *Seeing Like a State*, 327.

8. Miles, *King of the Wildcatters*, 15–16.

9. *Kansas City Star*, May 5, 1929; Glasscock, *Then Came Oil*, 218.

10. Glasscock, *Then Came Oil*, 218.

11. Glasscock, *Then Came Oil*, 218.

12. Miles, *King of the Wildcatters*, 6–7.

13. Miles, *King of the Wildcatters*, 7.

14. Miles, *King of the Wildcatters*, 74–75.

15. Miles, *King of the Wildcatters*, 74–75.

16. Miles, *King of the Wildcatters*, 74–75; Owen, *Trek of the Oil Finders*, 540.

17. Miles, *King of the Wildcatters*, 80.

18. *Kansas City Star*, May 5, 1929.

19. Another wildcatter expressed clearly how people instinctively attributed

the successful application of the belt-line theory to magic before geologists began offering scientific rationales: "From some unknown reason there was a belief that there was some magic in the 96th meridian. . . . At that time anything east of Dewey was taboo, but not for any geological reasons" (John H. Kane to James A. Veasey, May 17, 1941, "Belt-line Theory," box 23, JVC).

20. *Kansas City Star*, May 5, 1929.

21. *Kansas City Star*, May 5, 1929.

22. *Kansas City Star*, May 5, 1929.

23. Tom Slick Jr., "Some Comments on the Life of Tom Slick, Sr.," presented on the occasion of a testimonial award at the Cushing Petroleum Festival, September 9, 1952, box 4, file 6, Ray Miles Collection, Western History Collection, University of Oklahoma.

24. Slick, "Some Comments."

25. Tom Slick Jr., interview by Allan Nevins, July, 30 1951, Benedum and the Oil Industry Collection, Oral History Collection, Columbia University, New York; photocopy in box 6, file 4, Ray Miles Collection, Western History Collection, University of Oklahoma.

26. Tait, *Wildcatters*, 127.

27. The Bartlesville discovery well was drilled in 1897. For a full account, see Glasscock, *Then Came Oil*.

28. James A. Veasey to Alf M. Landon, June 12, 1941, "Belt-line Theory," box 23, JVC.

29. James A. Veasey to George Otis Smith, October 24, 1941, "Belt-line Theory," box 23, JVC.

30. J. S. Sidwell to James A. Veasey, July 14, 1941, "Belt-line Theory," box 23, JVC.

31. Veasey to Smith, October 24, 1941.

32. Sidwell to Veasey, July 14, 1941.

33. James A. Veasey to W. E. Wrather, June 26, 1941, "Belt-line Theory," box 23, JVC.

34. Veasey to Wrather, June 26, 1941. Veasey explained to Alf M. Landon that oil men who came from the East "brought the belt-line theory with them" (Veasey to Landon, June 12, 1941).

35. Alf Landon to James A. Veasey, June 18, 1941, "Belt-line Theory," box 23, JVC.

36. Landon to Veasey, June 18, 1941.

37. Landon to Veasey, June 18, 1941.

38. Landon to Veasey, June 18, 1941.

39. Owen, *Trek of the Oil Finders*, 293.

40. DeGolyer, *Development of the Art of Prospecting*, 27.

41. Arnold, "Two Decades of Petroleum Geology," 613.

42. Arnold, "Two Decades of Petroleum Geology," 613.

43. Among the list of greats, DeGolyer included John H. Galey, Mike Benedum, Edward L. Doheney, and Tom Slick.

44. Everette DeGolyer to Wallace E. Pratt, December 16, 1954, box 12, file 1513, EDC.

45. DeGolyer to Pratt, December 16, 1954.

46. DeGolyer to Pratt, December 16, 1954.

47. Wallace Pratt, *Oil in the Earth*, 58.

48. Wallace Pratt, *Oil in the Earth*, 57.

49. Wallace E. Pratt to Everette DeGolyer, January 24, 1955, box 12, file 1513, EDC.

50. DeGolyer to Pratt, December 16, 1954.

51. *Literary Digest*, March 14, 1914, 568.

52. "Hunch, Backed by Unfailing Courage," 32.

53. "Hunch, Backed by Unfailing Courage," 32.

54. "Hunch, Backed by Unfailing Courage," 32.

55. "Hunch, Backed by Unfailing Courage," 32.

56. "Hunch, Backed by Unfailing Courage," 32.

57. Scott, *Seeing Like a State*, 313.

58. Scott, *Seeing Like a State*, 312.

59. Scott, *Seeing Like a State*, 316.

60. "Metis knowledge is often so implicit and automatic that its bearer is at a loss to explain it" (Scott, *Seeing Like a State*, 329).

61. DeGolyer, Foreword, x.

62. DeGolyer, Foreword, x.

63. DeGolyer, Foreword, x.

64. Wallace E. Pratt to Everette DeGolyer, June 20, 1941, box 12, file 1513, EDC.

65. Scott, *Seeing Like a State*, 329.

66. Scott, *Seeing Like a State*, 329.

67. Forest Rees, circular letter mailed to oil men, General Correspondence 1919, Charles Decker Collection, Western History Collection, University of Oklahoma.

68. Rees, circular letter.

69. Rees, circular letter.

70. Scott, *Seeing Like a State*, 319.

71. Scott, *Seeing Like a State*, 320.

72. Scott, *Seeing Like a State*, 320.

73. Scott, *Seeing Like a State*, 317.

74. DeGolyer, *Development of the Art of Prospecting*, 25. For other examples, see DeGolyer to Pratt, February 16, 1955, and December 16, 1954, box 12, file 1513, EDC.

75. DeGolyer to Pratt, December 16, 1954.

76. DeGolyer to Pratt, December 16, 1954.

77. DeGolyer, Foreword, viii.

78. DeGolyer, Foreword, viii.

79. DeGolyer, Foreword, ix.

80. DeGolyer, Foreword, ix.

81. DeGolyer, *Development of the Art of Prospecting*, 27.

82. DeGolyer, *Development of the Art of Prospecting*, 27.

83. DeGolyer, *Development of the Art of Prospecting*, 27.

84. DeGolyer, *Development of the Art of Prospecting*, 27.

85. Everette DeGolyer to Wallace E. Pratt, December 15, 1954, box 12, file 1513, EDC. Note that even though DeGolyer is making the statement, he attributes the idea to Pratt.

86. DeGolyer, *Development of the Art of Prospecting*, 27.

87. DeGolyer, *Development of the Art of Prospecting*, 27.

88. Scott, *Seeing Like a State*, 324.

89. Scott, *Seeing Like a State*, 324.

90. Miles, *King of the Wildcatters*, 24.

91. *Literary Digest*, March 14, 1914, 568.

92. For a more detailed description of the events leading up to the discovery of Cushing and all the major figures involved, see Tyson, Thomas, and Faulk, *The McMan*; Knowles, *Greatest Gamblers*; Glasscock, *Then Came Oil*, 217–25; Tait, *Wildcatters*, 129; Owen, *Trek of the Oil Finders*, 294; Rister, *Oil*, 68–78.

93. Tait, *Wildcatters*, 129.

94. Jesse V. Howell, "History of Petroleum Geology," 3–4, box 16, Jesse V. Howell Collection, American Heritage Center, University of Wyoming. Eventually, oil men would continue drilling even deeper. Homer F. Wilcox missed the shallow sands entirely when in 1914 he sank a well southeast of Tulsa. This layer soon became the state's primary producing layer of sand and was forever known as the Wilcox sands. See Tait, *Wildcatters*, 129.

95. Powers, "Petroleum Geology in Oklahoma," 8.

96. Powers, "Petroleum Geology in Oklahoma," 8; Tait, *Wildcatters*, 129.

97. Rister, *Oil*, 124.

98. Owen, *Trek of the Oil Finders*, 294.

99. Powers, "Petroleum Geology in Oklahoma," 8.

100. Powers, "Petroleum Geology in Oklahoma," 8.

101. Howell, "History of Petroleum Geology," 20–21; Owen, *Trek of the Oil Finders*, 294.

102. Howell, "History of Petroleum Geology," 21.

103. Howell, "History of Petroleum Geology," 21.

104. Everette DeGolyer to James A. Veasey, June 10, 1941, "Belt-line Theory," box 23, JVC.

105. Owen, *Trek of the Oil Finders*, 294.

106. Howell, "History of Petroleum Geology," 8.

107. James O. Lewis to James A. Veasey, September 10, 1941, "Belt-line Theory," box 23, JVC.

108. DeGolyer felt that although Cushing was important, development of Augusta and El Dorado provided "the greatest impulse" toward acceptance of the anticlinal theory. DeGolyer to Veasey, June 10, 1941. Wallace Wrather, a petroleum geologist and one-time director of the USGS, agreed that Cushing played a significant role in demonstrating "that oil accumulation is governed by structure," and he agreed that Augusta and El Dorado illustrated the same point. I will cover these oil fields in a subsequent chapter. W. E. Wrather to James A. Veasey, November 3, 1941, "Belt-line Theory," box 23, JVC.

4. INSTITUTIONAL AUTHORITY

1. Bruce, *Launching of Modern American Science*, 200.

2. Lucier, *Scientists and Swindlers*, 322–23.

3. Lucier, *Scientists and Swindlers*, 323.

4. Bender, Hall, Haskell, and Mattingly, "Institutionalization and Education," 452–53.

5. Bender, "Cultures of Intellectual Life," 190.

6. Bender, "Cultures of Intellectual Life," 187.

7. Bender, Hall, Haskell, and Mattingly, "Institutionalization and Education," 463.

8. Bender, "Cultures of Intellectual Life," 187.

9. Bender, "Cultures of Intellectual Life," 187.

10. Gould, *Travels through Oklahoma*, 68.

11. Gould, *Covered Wagon Geologist*, 68.

12. Veysey, *Emergence of the American University*, 317.

13. Levy, *University of Oklahoma*, 3.

14. Gould, "Beginning of the Geological Work in Oklahoma," 200.

15. Gould, "Beginning of the Geological Work in Oklahoma," 199. Total student enrollment was somewhat higher than sixty because the university functioned as a preparatory school for a number of years and counted high school students among its population.

16. Gould, "Beginning of the Geological Work in Oklahoma," 200.

17. Gould, "Beginning of the Geological Work in Oklahoma," 200.

18. Gould, "Beginning of the Geological Work in Oklahoma," 200, and "Pioneer Geology in Oklahoma," 56.

19. Owen, *Trek of the Oil Finders*, 188; Howell, "History of Petroleum Geology."

20. "Its Geological Survey: The Party from the University Is Now in Logan and Noble Counties," *Kansas City Star*, June 19, 1900, box 1, folder 11, Sardis Roy Hadsell Collection, Western History Collection, University of Oklahoma.

21. "Its Geological Survey," *Kansas City Star*.

22. Gould, "Billion Barrels of Oil."

23. Gould, Hutchison, and Gaylord, *Preliminary Report*; Gould, "Billion Barrels of Oil."

24. Gould, "Petroleum and Natural Gas," 791.

25. Gould, "Oklahoma: An Example of Arrested Development," 426, 450.

26. Gould, "Oklahoma: An Example of Arrested Development," 436.

27. Beebe, Introduction.

28. Gould, *Covered Wagon Geologist*, 111.

29. Everette DeGolyer to James A. Veasey, June 10, 1941, "Belt-line Theory," JVC.

30. Gould, *Oklahoma: The Geologist's Laboratory*, 3.

31. Gould, *Oklahoma: The Geologist's Laboratory*, 3.

32. Gould, *Oklahoma: The Geologist's Laboratory*, 3.

33. Huffman, *History of the School of Geology and Geophysics*, 37–38.

34. Gould, "Pioneer Geology in Oklahoma," 56.

35. Gould, *Covered Wagon Geologist*, 113.

36. Gould, *Covered Wagon Geologist*, 115.

37. Gould, *Covered Wagon Geologist*, 112.

38. Gould, *Oklahoma: The Geologist's Laboratory*, 11.

39. Gould, *Covered Wagon Geologist*, 112.

40. Gould and Lewis, *The Permian of Western Oklahoma*, 11.

41. Gould, *Covered Wagon Geologist*, 112–13.

42. Aldrich, "Women in Paleontology."

43. Gould, *Covered Wagon Geologist*, 115.

44. Gould, *Covered Wagon Geologist*, 115.

45. Gould, *Oklahoma: The Geologist's Laboratory*, 4.

46. Horace L. Griley, "General Information," Horace L. Griley Letters, box 3A, EOC.

47. Everett Carpenter to Edgar Owen, August 3, 1963, box 1A, EOC.

48. Carpenter to Owen, August 3, 1963.

49. Beebe, Introduction, viii.

50. Beebe, Introduction, viii.

51. Beebe, Introduction, viii.

52. Beebe, Introduction, viii.

53. Beebe, Introduction, viii.

54. Kuhn, *Structure of Scientific Revolutions*, 111.

55. Kuhn, *Structure of Scientific Revolutions*, 111.

56. Kuhn, *Structure of Scientific Revolutions*, 846.

57. Gould, *Covered Wagon Geologist*, 171.

58. Gould, *Covered Wagon Geologist*, 191.

59. For the result of this field expedition, see Gould, *Geology and Water Resources*.

60. Gould, *Covered Wagon Geologist*, 191.

61. Gould, *Covered Wagon Geologist*, 191.

62. Gould, *Covered Wagon Geologist*, 191–92.

63. Hager, "Evidence of the Oklahoma Oil Fields."

64. Howell, "Historical Development of the Structural Theory," 20.

65. Goodrich, "Past and the Future," 451.

66. Goodrich, "Past and the Future," 452.

67. Goodrich, "Past and the Future," 452.

68. Goodrich, "Past and the Future," 454.

69. Dupree, *Science in the Federal Government*; Manning, *Government in Science*.

70. Manning, *Government in Science*, 218.

71. Manning, *Government in Science*, 93–94.

72. Manning, *Government in Science*, 209.

73. Carroll H. Wegemann to James A. Veasey, September 9, 1941, "Belt-line Theory," box 23, JVC.

74. Wegemann to Veasey, September 9, 1941.

75. Wegemann to Veasey, September 9, 1941.

76. E. Russell Lloyd to James A. Veasey, October 24, 1941, "Belt-line Theory," box 23, JVC.

77. Lloyd to Veasey, October 24, 1941.

78. Wegemann to Veasey, September 9, 1941.

79. DeGolyer, *Development of the Art of Prospecting*, 26.

80. Eldridge fell ill and was unable to complete the report. After his death Ralph Arnold took over the task of preparing the manuscript, publishing it in 1907 as USGS Bulletin number 309. See Hamilton, "Development of Petroleum Geology in California," 457–58.

81. Hamilton, "Development of Petroleum Geology in California," 458.

82. Hamilton, "Development of Petroleum Geology in California," 458.

83. Wegemann to Veasey, September 9, 1941.

84. Rabbitt and Rabbitt, "U.S. Geological Survey," 751; Rabbitt, *United States Geological Survey*, 22–24.

85. Hendrickson, "Nineteenth-Century State Geological Surveys," 371.

86. Gould, "Geological Work in the Southwest," 32.

87. Gould, *Covered Wagon Geologist*, 49.

88. Gould, "Beginning of the Geological Work in Oklahoma," 201.

89. For a lengthy description of the journey and a map illustrating the exact route, see Gould, *Covered Wagon Geologist*, 117–28, 130.

90. The three publications are USGS water-supply papers 148, 154, and 191. For a full citation of each, see Clifton, "Memorial to Charles Newton Gould," 172. The USGS and Oklahoma Geological Survey (OGS) cooperated in field work performed in the Wichita and Arbuckle mountains and in the red beds region of western Oklahoma. The OGS published a two-volume report on petroleum and natural gas in the state that included several reports written by USGS geologists. See Howell, "History of Petroleum Geology." The federal and state surveys collaborated again in 1927 to produce a colored map that Gould bragged was the most complete and comprehensive state geological map ever published in the United States. See Gould, *Oklahoma: The Geologist's Laboratory*, 6.

91. Gould, *Covered Wagon Geologist*, 168.

92. Gould, *Covered Wagon Geologist*, 168.

93. Gould, *Covered Wagon Geologist*, 168.

94. Gould, *Covered Wagon Geologist*, 168.

95. Gould, *Covered Wagon Geologist*, 168.

96. Gould, "Geological Work in the Southwest," 32.

97. Gould, *Covered Wagon Geologist*, 141.

98. Gould, *Covered Wagon Geologist*, 141.

99. Gould, *Covered Wagon Geologist*, 141.

100. Gould, *Travels through Oklahoma*, 142.

101. Gould, *Covered Wagon Geologist*, 142.

102. Gould, *Covered Wagon Geologist*, 142.

103. Gould, *Covered Wagon Geologist*, 142.

104. Gould, *Covered Wagon Geologist*, 143.

105. Gould, *Covered Wagon Geologist*, 143.

106. Howell, "History of Petroleum Geology." Howell names approximately twenty men who worked for the survey, many of whom went on to distinguished careers and are mentioned throughout my study. For information similar to Howell's, see Owen, *Trek of the Oil Finders*, 278. Note that Owen almost certainly derived his information from the Howell document.

107. Gould, *Covered Wagon Geologist*, 147.

108. Gould, *Covered Wagon Geologist*, 147.

109. Charles N. Gould to Charles W. Brown, June 1, 1908, box M2179, file 1-1, L. L. Hutchison Collection, Western History Collection, University of Oklahoma.

110. Gould, "Memorial: Lon Lewis Hutchison."

111. Gould, "Memorial: Lon Lewis Hutchison."

112. Gould to Brown, June 1, 1908.

113. Gould, "Memorial: Lon Lewis Hutchison."

114. Owen, *Trek of the Oil Finders*, 293.

115. Gould, "Memorial: Lon Lewis Hutchison," 651.

116. L. L. Hutchison to Tom Wall, February 14, 1911, box 1, folder 2, L. L. Hutchison Collection, Western History Collection, University of Oklahoma.

117. Owen, *Trek of the Oil Finders*, 294.

118. Glasscock, *Then Came Oil*, 315. Buttram's daughter wrote a very filiopietistic biography of her father, but it is useful for factual details. See Buttram and Fleming, *One Man's Footprints*; Branson, "Frank Buttram," 220.

119. Glasscock, *Then Came Oil*, 315–16.

120. Glasscock, *Then Came Oil*, 317.

121. Glasscock, *Then Came Oil*, 317.

122. Glasscock, *Then Came Oil*, 317.

123. Glasscock, *Then Came Oil*, 220.

124. Howell, "History of Petroleum Geology."

125. Howell, "History of Petroleum Geology," 8; Buttram, *Cushing Oil and Gas Field*.

126. Powers, "Petroleum Geology in Oklahoma," 8; Buttram and Fleming, *One Man's Footprints*, 35.

127. Bloesch, "Early Day Petroleum Geology in Oklahoma," box 3A, file "Ed Bloesch Letters," EOC.

128. Powers, "Petroleum Geology in Oklahoma," 8.

129. Powers, "Petroleum Geology in Oklahoma," 8.

130. Powers, "Petroleum Geology in Oklahoma," 8.

131. Quoted in Buttram and Fleming, *One Man's Footprints*, 41.

132. Newby, "Daniel Webster Ohern."

133. Buttram and Fleming, *One Man's Footprints*, 13. For a more complete description of the location of the wells they discovered, see Newby, "Daniel Webster Ohern," 963.

134. Newby, "Daniel Webster Ohern," 963.

135. Buttram and Fleming, *One Man's Footprints*, 15.

136. Howell, "History of Petroleum Geology," 8–9; Buttram and Fleming, *One Man's Footprints*, 15.

137. Newby, "George Franklin Buttram," 245. Buttram's daughter says that Fortuna was sold for six million dollars. See Buttram and Fleming, *One Man's Footprints*, 13.

138. Buttram and Fleming, *One Man's Footprints*, 43.

139. DeGolyer, *Development of the Art of Prospecting*, 17.

140. Clapp, "Studies in the Application of the Anticlinal Theory," 565–66; Lahee, "Education of the Geologist."

141. Gould, *Covered Wagon Geologist*, xii.

1. Most of the literature on the organizational synthesis makes this point. For examples, see Galambos, "Technology, Political Economy, and Professionalization," 471.

2. Standard did find a way to access the region through its multiple subsidiaries, but the Supreme Court ruling greatly complicated its ability to function in its traditional monopolistic manner. See Joseph Pratt, "Petroleum Industry in Transition."

3. I borrow this term from Hughes, "Evolution of Large Technological Systems."

4. I specifically use the term to connote business as well as areas outside business such as politics and science. See Hughes, *American Genesis*, 3. Alfred Chandler casts the term "system-building" as solely a "managerial" tactic of "organization" and "coordination." See his chap. 5, "System-Building, 1880s–1900," in *Visible Hand*.

5. Hughes, *American Genesis*, 3.

6. Hughes, *Networks of Power*.

7. Carpenter, "Reminiscences of Everett Carpenter," 431.

8. Doherty, *Principles and Ideas for Doherty Men*, 143.

9. Jones, *Cities Service Story*, 22.

10. Jones, *Cities Service Story*, 22.

11. Jones, *Cities Service Story*, 22.

12. Doherty, *Principles and Ideas for Doherty Men*, 104.

13. Doherty, *Principles and Ideas for Doherty Men*, 106.

14. Doherty, *Principles and Ideas for Doherty Men*, 105.

15. Doherty, *Principles and Ideas for Doherty Men*, 105–7.

16. Doherty, *Principles and Ideas for Doherty Men*, 104.

17. Doherty, *Principles and Ideas for Doherty Men*, 104.

18. Doherty, *Principles and Ideas for Doherty Men*, 104.

19. Doherty, *Principles and Ideas for Doherty Men*, 107. This was not an uncommon feeling, as field engineers in other professions typically resented coworkers who had received theoretical training. See Noble, *America by Design*, 27.

20. Doherty, *Principles and Ideas for Doherty Men*, 106–7.

21. Doherty, *Principles and Ideas for Doherty Men*, 105.

22. Doherty, *Principles and Ideas for Doherty Men*, 105.

23. Doherty, *Principles and Ideas for Doherty Men*, 105.

24. Doherty, *Principles and Ideas for Doherty Men*, 105.

25. Doherty, *Principles and Ideas for Doherty Men*, 104.

26. Doherty, *Principles and Ideas for Doherty Men*, 104.

27. Rose, *Cities of Light and Heat*, 74. Doherty's inability to find suitably trained engineers was a common problem for other industries as well, and his solution

of establishing an in-house training problem was also a common solution. See Noble, *America by Design*, 29.

28. Rose, *Cities of Light and Heat*, 74.

29. Rose, *Cities of Light and Heat*, 74.

30. Rose, *Cities of Light and Heat*, 74, 75; W. Alton Jones, "Services of Henry L. Doherty and Company to Subsidiaries of Cities Service Company," 6, vol. II, chap. 6, box 14, Cities Service Collection, Western History Collection, University of Oklahoma (hereafter cited as CSC).

31. Jones, "Services of Henry L. Doherty," 7–8.

32. Jones, "Services of Henry L. Doherty," 6.

33. Jones, "Services of Henry L. Doherty," 6.

34. Jones, "Services of Henry L. Doherty," 9.

35. Jones, "Services of Henry L. Doherty," 7.

36. Jones, "Services of Henry L. Doherty," 7.

37. Warner, "Sources of Men," 55.

38. Jones, "Services of Henry L. Doherty," 8.

39. Jones, "Services of Henry L. Doherty," 8. The total number of students who received training in the oil and gas business alone equaled 352 from 1916 to 1931.

40. Jones, "Services of Henry L. Doherty," 11.

41. Jones, "Services of Henry L. Doherty," 11.

42. Ellis, *On the Oil Lands*, 40, 55. According to Ellis, Doherty acquired a total of eighteen companies on July 1, 1912. See *On the Oil Lands*, 56n8 for a list of all but two of these. However, Everett Carpenter, who worked for Empire, reported that Doherty acquired from Barnsdall "about fifty-six separate oil properties and two gas companies" (Everett Carpenter, "As I Remember It," *Shale Shaker*, June 1957, 40, box 5, folder "Everett Carpenter — Company's 1st Full-time Geologist," box 5, CSC).

43. Owen, *Trek of the Oil Finders*, 298. Although Doherty acquired Barnsdall's shares in ITIO, he did not become the majority shareholder until later.

44. Karasick and Karasick, *Oilman's Daughter*, 5; *Derrick's Hand-Book of Petroleum*, 1:870.

45. Karasick and Karasick, *Oilman's Daughter*, 6.

46. Karasick and Karasick, *Oilman's Daughter*, 6.

47. Karasick and Karasick, *Oilman's Daughter*, 6.

48. Hanson, "Senator William B. Pine," 6.

49. Hanson, "Senator William B. Pine," 6.

50. Hanson, "Senator William B. Pine," 6.

51. Hanson, "Senator William B. Pine," 9.

52. Hanson, "Senator William B. Pine," 10.

53. Hanson, "Senator William B. Pine," 10.

54. Lois Straight Johnson, "Bartlesville Historical Society," April 16, 1975, H. R. Straight, box 14, CSC.

55. Lois Johnson, "Bartlesville Historical Society."

56. Lois Johnson, "Bartlesville Historical Society."

57. Lois Johnson, "Bartlesville Historical Society." Straight's reasons for attending college are not entirely clear, but one latter-day biographical sketch offered the vague statement that he "began to realize the importance and need for academic background" and thus enrolled at Stanford. See "Biographical Sketch — Herbert R. Straight," October 8, 1954, H. R. Straight, box 14, CSC; Lois Johnson, "Bartlesville Historical Society." This file contains numerous but brief biographies of Straight that mostly take the form of press releases.

58. Lois Johnson, "Bartlesville Historical Society"; Herbert Straight to James A. Veasey, May 17, 1941, "Belt-line Theory," box 23, JVC.

59. Although a practical oil man, Barnsdall probably had no aversion to geology, and it may have even appealed to him and motivated him to hire both Pine and Straight. Petroleum geology was accepted by oil companies outside the United States at a much earlier date. The first company to employ geology in Oklahoma was the Union des Pétroles d'Oklahoma in 1911; this company originated out of Holland and Paris. Straight worked for this company at some point, but I have been unable to determine the beginning date of his employment. It is highly possible that this position drew him to Oklahoma, and the geology he practiced there made him an even more attractive candidate to Barnsdall and eventually to Doherty. See Owen, *Trek of the Oil Finders*, 291–92; Ellison, Jones, and Owen, *Flavor of Ed Owen*, 11.

60. Everett Carpenter to Edgar W. Owen, July 31, 1963, box 1, folder 88, Everett Carpenter Collection, Western History Collection, University of Oklahoma. This letter is reprinted in its entirety as "Reminiscences of Everett Carpenter," 431.

61. They hired petroleum geologist Lee Hager in 1901. See Lee Hager to James A. Veasey, June 5, 1941, "Belt-line Theory," box 23, JVC.

62. Owen, *Trek of the Oil Finders*, 298.

63. Owen, *Trek of the Oil Finders*, 298.

64. Arthur B. Fox, "The Incline Builders: Forgotten Engineers of Pittsburgh," *Pittsburgh Tribune-Review*, June 1, 1997.

65. Fox, "The Incline Builders"; Carpenter to Owen, July 31, 1963.

66. Carpenter to Owen, July 31, 1963.

67. Carpenter to Owen, July 31, 1963.

68. Carpenter to Owen, July 31, 1963.

69. John H. Kane to James A. Veasey, July 15, 1941, "Belt-line Theory," box 23, JVC.

70. Carpenter to Owen, July 31, 1963.

71. Carpenter to Owen, July 31, 1963.

72. Carpenter to Owen, July 31, 1963.

73. Carpenter to Owen, July 31, 1963.

74. Carpenter to Owen, July 31, 1963.

75. Carpenter to Owen, July 31, 1963.

76. Carpenter to Owen, July 31, 1963.

77. Carpenter to Owen, July 31, 1963.

78. Carpenter to Owen, July 31, 1963.

79. Carpenter to Owen, July 31, 1963.

80. Carpenter to Owen, July 31, 1963.

81. Carpenter to Owen, July 31, 1963.

82. Carpenter to Owen, July 31, 1963.

83. Carpenter to Owen, July 31, 1963.

84. Carpenter to Owen, July 31, 1963.

85. Carpenter to Owen, July 31, 1963.

86. Carpenter to Owen, July 31, 1963.

87. Carpenter to Owen, July 31, 1963.

88. Horace L. Griley to Mr. Jenkins, March 4, 1973, Horace L. Griley Letters, box 3A, EOC. A clear consensus exists among geologists at the time that Empire's discovery of Augusta and El Dorado marked a turning point in their profession. For examples, see the following: William N. Davis to James Veasey, June 5, 1941, "Belt-line Theory," box 23, JVC.

89. Ellison, Jones, and Owen, *Flavor of Ed Owen*, 21.

90. Ellison, Jones, and Owen, *Flavor of Ed Owen*, 21.

91. Carpenter to Owen, July 31, 1963.

92. Alex W. McCoy to James A. Veasey, June 20, 1941, "Belt-line Theory," box 23, JVC.

93. L. Murray Nuemann to James Veasey, May 20, 1941, "Belt-line Theory," box 23, JVC.

94. Horace L. Griley to Edgar W. Owen, September 14, 1965, Empire Gas and Fuel Co. (Cities Service), box 3A, EOC. Elsewhere, Griley states that the number was closer to two hundred. See Griley to Jenkins, March 4, 1973.

95. Carpenter reported, "I am unable to arrive at so large a figure" as 250. He continued, "I have been unable to arrive at a number greater than a hundred, but there may have been several whose names have been overlooked" (Carpenter to Owen, July 31, 1963). Owen attempted to document every geologist Empire employed from 1916 to 1919, and he compiled 212 names plus 8 "probables" for a total of 220, although they probably were not all employed at the same time. See "Empire Gas and Fuel Co. Geology Department Personnel," Empire Gas and

Fuel Co. (Cities Service), box 3A, EOC. John Steiger claimed that the geological staff grew "to almost 250 by 1917" (John Steiger, untitled document, June 8, 1968, miscellaneous text documents, box 11, CSC). See also Lois Johnson, "Bartlesville Historical Society." Perhaps the most definitive source is that which says, "Some records which we have found indicate a total of 149 geologists were on the pay roll during the summer of 1917" (Straight to Veasey, May 17, 1941).

96. Roswell Johnson to James A. Veasey, February 24, 1941, "Belt-line Theory," box 23, JVC.

97. Owen, *Trek of the Oil Finders*, 225. For a more detailed description of the instruments and techniques employed at the time, see Owen, *Trek of the Oil Finders*, 225–27, 241–42, 295–97, and the following sources: Hayes, *Handbook for Field Geologists* and *Field Mapping for the Oil Geologist*; Lahee, *Field Geology*; English, "Some Planetable Methods," 47–54.

98. Owen, *Trek of the Oil Finders*, 225.

99. Owen, *Trek of the Oil Finders*, 225–26.

100. Carpenter to Owen, July 31, 1963.

101. "Scientific Methods of Exploration," 14.

102. J. W. George to W. A. Tarr, October 30, 1917, Oklahoma Notes and Extracts, Pre-1921, box 1C, EOC.

103. Roswell Johnson to Veasey, February 24, 1941.

104. Ellison, Jones, and Owen, *Flavor of Ed Owen*, 21.

105. Ellison, Jones, and Owen, *Flavor of Ed Owen*, 21.

106. "Scientific Methods of Exploration," 14.

107. DeGolyer, "Concepts on Occurrence of Oil and Gas," 24.

108. DeGolyer, "Concepts on Occurrence of Oil and Gas," 24.

109. Wallace Pratt, *Oil in the Earth*, 23.

110. Wallace Pratt, *Oil in the Earth*, 23.

111. Carpenter to Owen, July 31, 1963.

112. Carpenter to Owen, July 31, 1963.

113. Everett Carpenter to Charles N. Gould, December 19, 1924, quoted in Branson, "Petroleum Notes from the Twenties," 94.

114. Carpenter to Gould, December 19, 1924.

115. Carpenter to Gould, December 19, 1924.

116. Carpenter to Gould, December 19, 1924. For further explanation of shoe-string and lense traps, see Levorsen, *Geology of Petroleum*, 293–318.

117. Branson, "Petroleum Notes from the Twenties," 94.

118. Ball, Ball, and Tuner, *This Fascinating Oil Business*, 37. For further elaboration on stratigraphic traps, see Landes, *Petroleum Geology*, 276–79, 388–91, and Levorsen, *Geology of Petroleum*, 287–339.

119. Ball, Ball, and Turner, *This Fascinating Oil Business*, 39–40.

120. Levorsen, "Stratigraphic Versus Structural Accumulation," 524.

121. Warner, "Sources of Men," 38. An equally, if not more, viable candidate is Edwin T. Dumble, who ran the geological department for the Southern Pacific Railroad in California, but this debate is beyond the scope of my present purpose. See Dorsey Hager to James A. Veasey, June 18, 1941, "Belt-line Theory," box 23, JVC.

122. Owen, *Trek of the Oil Finders*, 107.

123. Owen, *Trek of the Oil Finders*, 108.

124. The history of subsurface geology unfolded over a number of years and involved a number of geologists. For the best source for identifying the geologists who made the greatest contributions and for a description of their work, see Owen, *Trek of the Oil Finders*, 106–16, 127–33, 166–73.

125. Executive Committee, "Memorial: Alexander Watts McCoy," 293.

126. Landes, *Petroleum Geology*, 84.

127. Landes, *Petroleum Geology*, 84.

128. McCoy to Veasey, June 20, 1941.

129. Alex W. McCoy to James A. Veasey, July 15, 1941, "Belt-line Theory," box 23, JVC. For another example of McCoy's belief that his efforts constituted "the beginning in the Midcontinent of petroleum engineering as it is known today," see the same file for McCoy to Veasey, June 20, 1941.

130. McCoy to Veasey, June 20, 1941.

131. McCoy to Veasey, July 15, 1941.

132. McCoy to Veasey, July 15, 1941.

133. McCoy to Veasey, July 15, 1941.

134. McCoy to Veasey, July 15, 1941.

CONCLUSION

1. "I. W. W. Plot Breaks Prematurely in Blowing Up of Pew Residence," *Tulsa World*, October 30, 1917. For more on this incident and the impact of the Wobblies on the oil industry generally, see Sellars, *Oil, Wheat, and Wobblies*.

2. "I. W. W. Plot Breaks Prematurely," *Tulsa World*, and Sellars, *Oil, Wheat, and Wobblies*.

3. "I. W. W. Plot Breaks Prematurely," *Tulsa World*, and Sellars, *Oil, Wheat, and Wobblies*.

4. Richard White, "From Wilderness to Hybrid Landscapes," 560–61; Jacoby, *Crimes against Nature*; Warren, *Hunter's Game*.

5. Richard White, "'Are You an Environmentalist?,'" 173.

6. Richard White, "'Are You an Environmentalist?,'" 173.

7. *Kansas City Star*, May 5, 1929.

Bibliography

ARCHIVAL SOURCES

Charles Decker Collection. Western History Collection. University of Oklahoma.
Charles Gould Collection. Western History Collection. University of Oklahoma.
Cities Service Collection. Western History Collection. University of Oklahoma.
Earl Oliver Collection. American Heritage Center. University of Wyoming.
Edgar Wesley Owen Collection. American Heritage Center. University of Wyoming.
Everett Carpenter Collection. Western History Collection. University of Oklahoma.
Everette DeGolyer Collection. Southern Methodist University.
George Otis Smith Collection. American Heritage Center. University of Wyoming.
James A. Veasey Collection. American Heritage Center. University of Wyoming.
Jesse V. Howell Collection. American Heritage Center. University of Wyoming.
J. P. Lesley Collection. American Philosophical Society. Philadelphia.
L. L. Hutchison Collection. Western History Collection. University of Oklahoma.
Oklahoma Geological Survey Collection. Western History Collection. University
    of Oklahoma.
Oral History of the Oil Industry Collection. Center for American History. Austin TX.
Ray Miles Collection. Western History Collection. University of Oklahoma.
Sardis Roy Hadsell Collection. Western History Collection. University of Oklahoma.
Sun Oil Collection. Hagley Museum and Library. Wilmington DE.

PUBLISHED WORKS

Abbott, Andrew. *The System of Professions: An Essay on the Division of Expert Labor.*
    Chicago: University of Chicago Press, 1988.
Adams, Sean P. *Old Dominion, Industrial Commonwealth: Coal, Politics, and Economy
    in Antebellum America.* Baltimore: Johns Hopkins University Press, 2004.
———. "Partners in Geology, Brothers in Frustration." *Virginia Magazine of History & Biography* 106, no. 1 (1998): 5–34.

Aldrich, Michele. "American State Geological Surveys, 1820–1845." In *Two Hundred Years of Geology in America*, edited by Cecil Schneer, 133–44. Hanover: University of New Hampshire Press, 1979.

———. "Women in Paleontology in the United States, 1840–1960." *Earth Sciences History* 1, no. 1 (1982): 14–22.

Ames, Mary Lesley, ed. *Life and Letters of Peter and Susan Lesley*. 2 vols. New York: Putnam, 1909.

Anderson, Robert O. *Fundamentals of the Petroleum Industry*. Norman: University of Oklahoma Press, 1984.

Andrews, Thomas G. *Killing for Coal: America's Deadliest Labor War*. Cambridge MA: Harvard University Press, 2008.

Arnold, Ralph. "Two Decades of Petroleum Geology, 1903–1922." *American Association of Petroleum Geologists Bulletin* 8 (November–December 1923): 603–24.

Asbury, Herbert. *The Golden Flood*. New York: Knopf, 1942.

Ball, Max, Douglas Ball, and Daniel S. Turner, eds. *This Fascinating Oil Business*. Indianapolis: Bobbs-Merrill, 1940.

Barton, Donald C. "The Wigglestick." *American Association of Petroleum Geologists Bulletin* 10 (March 1926): 312–13.

Beebe, B. W. Introduction. In *Covered Wagon Geologist*, by Charles Newton Gould, v–x. Norman: University of Oklahoma Press, 1959.

Ben-David, Joseph. *The Scientist's Role in Society: A Comparative Study*. Englewood Cliffs NJ: Prentice-Hall, 1971.

Bender, Thomas. "The Cultures of Intellectual Life: The City and the Professions." In *New Directions in American Intellectual History*, edited by John Higham and Paul Keith, 181–95. Baltimore: Johns Hopkins University Press, 1979.

Bender, Thomas, Peter D. Hall, Thomas L. Haskell, and Paul H. Mattingly. "Institutionalization and Education in the Nineteenth and Twentieth Centuries." *History of Education Quarterly* 20, no. 4 (1980): 449–72.

Best, Gary Dean. *The Politics of American Individualism: Herbert Hoover in Transition, 1918–1921*. Westport CT: Greenwood Press, 1975.

Bijker, Wiebe E., Thomas P. Hughes, and Trevor J. Pinch. Introduction. In *The Social Construction of Technological Systems: New Directions in the Sociology and History of Technology*, edited by Wiebe E Bijker, Thomas P. Hughes, and Trevor J. Pinch, 1–8. Cambridge MA: MIT Press, 1987.

Bishop, C. E. "Vacation in Petrolia." *Our Boys and Girls*, October 1871.

Black, Brian. *Petrolia: The Landscape of America's First Oil Boom*. Creating the North American Landscape. Baltimore: Johns Hopkins University Press, 2000.

Blakey, Ellen Sue. *Oil on Their Shoes: Petroleum Geology to 1920*. Tulsa: American Association of Petroleum Geologists, 1985.

Bledstein, Burton J. *The Culture of Professionalism: The Middle Class and the Development of Higher Education in America.* New York: Norton, 1976.

Boatright, Mody C. *Folklore of the Oil Industry.* Dallas: Southern Methodist University Press, 1963.

Boatright, Mody C., and William A. Owens. *Tales from the Derrick Floor: A People's History of the Oil Industry.* Lincoln: University of Nebraska Press, 1970.

Bone, J. H. A. *Petroleum and Petroleum Wells.* Philadelphia: Lippincott, 1865.

Boscoe, Francis P. "'The Insanities of an Exalted Imagination': The Troubled First Geological Survey of Pennsylvania." *Pennsylvania Magazine of History & Biography* 127, no. 3 (2003): 291–308.

Bowker, Geoffrey C., and Susan Leigh Star. *Sorting Things Out: Classification and Its Consequences.* Inside Technology. Cambridge MA: MIT Press, 1999.

Branson, Carl C. "Frank Buttram, 1886–1966." In *History of the School of Geology and Geophysics, the University of Oklahoma*, edited by George Garrett Huffman. Norman: Alumni Advisory Council of the School of Geology and Geophysics, University of Oklahoma, 1990.

———. "Petroleum Notes from the Twenties." *Oklahoma Geology Notes* 17, no. 10 (October 1957): 93–94.

Brown, JoAnne. *The Definition of a Profession: The Authority of Metaphor in the History of Intelligence Testing, 1890–1930.* Princeton NJ: Princeton University Press, 1992.

———. "Profession." In *Companion to American Thought*, edited by Richard Wightman Fox and James T. Kloppenberg, 543–46. Oxford, UK: Blackwell, 1995.

Bruce, Robert V. *The Launching of Modern American Science.* Ithaca NY: Cornell University Press, 1988.

Buttram, Frank. *The Cushing Oil and Gas Field, Oklahoma.* Oklahoma Geological Survey bulletin no. 18. Norman: Oklahoma Geological Survey, 1914.

Buttram, Merle, and J. Landis Fleming. *One Man's Footprints: The Story of Frank Buttram.* Muskogee OK: Western Heritage Books, 1985.

Cahan, David. "Looking at Nineteenth-Century Science: An Introduction." In *From Natural Philosophy to the Sciences*, edited by David Cahan, 3–15. Chicago: University of Chicago Press, 2003.

Calvert, Monte. *The Mechanical Engineer in America, 1830–1910: Professional Cultures in Conflict.* Baltimore: Johns Hopkins University Press, 1967.

———. "The Search for Engineering Unity: The Professionalization of Special Interest." In *Building the Organizational Society: Essays on Associational Activities in Modern America*, edited by Jerry Israel, 42–54. New York: Free Press, 1972.

Campbell, M. R. "Historical Review of Theories Advanced by American Geologists to Account for the Origin and Accumulation of Oil." *Economic Geology* 6 (1911): 363–95.

Cannon, Susan Faye. *Science in Culture: The Early Victorian Period*. New York: Dawson and Science History Publications, 1978.

Carll, John F. *The Geology of the Oil Regions of Warren, Venango, Clarion, and Butler Counties*. Report of Progress, Second Geological Survey of Pennsylvania. Harrisburg PA: Board of Commissioners for the Second Geological Survey, 1880.

————. *Report of Progress in the Venango County District*. Harrisburg PA: Board of Commissioners for the Second Geological Survey, 1875.

Carlson, W. Bernard. *Innovation as a Social Process: Elihu Thomson and the Rise of General Electric, 1870–1900*. Cambridge, UK: Cambridge University Press, 1991.

Carpenter, Everett. "Reminiscences of Everett Carpenter." In *Digest V: A Compilation of Unaltered Geologic Papers from Shale Shaker, Vols. 15–17 (1964–67)*, 431–36. Oklahoma City: Times-Journal Publishing Company, 1968.

Chance, Henry M. "A Biographical Notice of J. Peter Lesley." *Proceedings of the American Philosophical Society* 45 (1906): i–ixv.

————. *The Northern Townships of Butler County*. Harrisburg: Board of Commissioners for the Second Geological Survey, 1879.

Chandler, Alfred. *The Visible Hand: The Managerial Revolution in American Business*. Cambridge MA: Belknap Press, 1977.

Chernow, Ron. *Titan: The Life of John D. Rockefeller, Sr.* New York: Vintage, 1998.

Clapp, Frederick G. "The Occurrence of Petroleum." In *A Handbook of the Petroleum Industry*, edited by David T. Day, 1–166. New York: Wiley, 1922.

————. "Studies in the Application of the Anticlinal Theory of Oil and Gas Accumulation." *Economic Geology* 4, no. 6 (1917): 565–70.

Clark, Blue. "The Beginning of Oil and Gas Conservation in Oklahoma, 1907–1931." *Chronicles of Oklahoma* 55 (1977–78): 375–91.

Clark, John G. *Energy and the Federal Government: Fossil Fuel Policies, 1900–1946*. Urbana: University of Illinois Press, 1987.

Clifton, Roland L. "Memorial to Charles Newton Gould." *Proceedings Volume of the Geological Society of America*, annual report for 1949 (June 1950): 165–74.

Cohen, Benjamin R. *Notes from the Ground: Science, Soil, and Society in the American Countryside*. New Haven CT: Yale University Press, 2009.

Collins, H. M. "The TEA Set: Tacit Knowledge and Scientific Networks." *Science Studies* 4 (1974): 165–85.

Cooper, Augustus S. *Genesis of Petroleum and Asphaltum in California*. California State Mining Bureau bulletin no. 16. Sacramento: Superintendent State Printing, 1899.

Copithorne, W. L. "From Doodlebug to Seismography." *The Lamp* 64 (1992): 43–47.

Czarniawska-Joerges, Barbara, and Tor Hernes. *Actor-Network Theory and Organizing*. Malmö: Copenhagen Business School Press, 2005.

Daniels, George H. *American Science in the Age of Jackson.* New York: Columbia University Press, 1968.

——. "Professionalization in American Science." *Isis* 58 (1967): 151–66.

——. "The Pure-Science Ideal and Democratic Culture." *Science* 156 (June 30, 1967): 1699–705

——. *Science in American Society.* New York: Knopf, 1971.

Davis, William Morris. "Biographical Memoir of J. Peter Lesley, 1819–1903." *Biographical Memoirs, National Academy of Sciences* 8 (1915): 152–240.

Day, David T. *A Handbook of the Petroleum Industry.* 2 vols. New York: Wiley, 1922.

DeGolyer, Everette. "Concepts on Occurrence of Oil and Gas." In *History of Petroleum Engineering,* 17–26. New York: American Petroleum Institute, 1961.

——. *The Development of the Art of Prospecting.* Princeton NJ: Guild of Brackett Lecturers, 1940.

——. Foreword. In *Oil! Titan of the Southwest,* by Carl Coke Rister, vii–xi. Norman: University of Oklahoma Press, 1949.

The Derrick's Hand-Book of Petroleum: A Complete Chronological and Statistical Review of Petroleum Developments during 1859 to 1898. 2 vols. Oil City PA: Derrick Publishing, 1898.

DeVore, Paul W. *Technology: An Introduction.* Worcester MA: Davis Publications, 1980.

Dickey, Parke A. "Oil Is Found with Ideas." *Tulsa Geological Society Digest* 26 (1958): 84–101.

——. "Pennsylvania's Contribution to Petroleum Geology." *American Association of Petroleum Geologists Bulletin* 73, no. 9 (1989): 1143.

Doherty, Henry L. *Principles and Ideas for Doherty Men,* compiled by Glenn Marston. Vol. 2. Privately printed for use by Doherty Organization members, 1923.

Dott, Robert H., Sr., and Merrill J. Reynolds, comp. *Sourcebook for Petroleum Geology.* Semicentennial Commemorative Volume, Memoir 5. Tulsa: American Association of Petroleum Geologists, 1969.

Dupree, A. Hunter. *Science in the Federal Government: A History of Policies and Activities to 1940.* Cambridge MA: Belknap Press, 1957.

Elkins, L. E. "Research." In *History of Petroleum Engineering,* 1085–1113. New York: American Petroleum Institute, 1961.

Ellis, William Donohue. *On the Oil Lands with Cities Service.* Tulsa: Cities Service Oil and Gas Corporation, 1983.

Ellison, Samuel P., Jr., Joseph J. Jones, and Mirva Owen, eds. *The Flavor of Ed Owen: A Geologist Looks Back.* Austin: Geology Foundation, University of Texas, 1987.

English, Walter A. "Some Planetable Methods." *American Association of Petroleum Geologists Bulletin* 8 (January–February 1924): 47–54.

Executive Committee. "Memorial: Alexander Watts McCoy." *American Association of Petroleum Geologists Bulletin* 30 (February 1946): 292–95.

Fabian, Ann. *Card Sharps, Dream Books, and Bucket Shops: Gambling in 19th-Century America*. Ithaca NY: Cornell University Press, 1990.

Fanning, Leonard. *The Story of the American Petroleum Institute*. Privately printed by the author, 1959.

Ferguson, Walter Keene. *Geology and Politics in Frontier Texas, 1845–1909*. Austin: University of Texas, 1967.

Fiege, Mark. *Irrigated Eden: The Making of an Agricultural Landscape in the American West*. Seattle: University of Washington Press, 1999.

Franks, Kenny. *Oklahoma Petroleum Industry*. Oklahoma City: Oklahoma Heritage Association, 1980.

———. *The Rush Begins: A History of the Red Fork, Cleveland, and Glenn Pool Oil Fields*. Oklahoma City: Western Heritage Books, 1984.

Fuller, Myron L. "Appalachian Oil Field." *Bulletin of the Geological Society of America* 28 (1917): 617–54.

Galambos, Louis. "Technology, Political Economy, and Professionalization: Central Themes of the Organizational Synthesis." *Business History Review* 57 (Winter 1983): 471–93.

Galbreath, Frank. *Glenn Pool: A Little Town of Yesteryear*. Privately printed by the author, 1978.

Galey, John T. "The Anticlinal Theory of Oil and Gas Accumulation: Its Role in the Inception of the Natural Gas and Modern Oil Industries in North America." In *Geologists and Ideas: A History of North America*, edited by Ellen T. Drake and William M. Jordan, 428–29. Boulder CO: Geological Society of America, 1985.

Geertz, Clifford. *Local Knowledge: Further Essays in Interpretive Anthropology*. New York: Basic Books, 1983.

Geison, Gerald L. *Professions and Professional Ideologies in America*. Chapel Hill: University of North Carolina Press, 1983.

Gerstner, Patty. *Henry Darwin Rogers, 1808–1866: American Geologist*. Tuscaloosa: University of Alabama Press, 1994.

Glasscock, Carl B. *Then Came Oil: The Story of the Last Frontier*. New York: Bobbs-Merrill, 1938.

Goetzmann, William. *Exploration and Empire: The Explorer and the Scientist in the Winning of the American West*. New York: Norton, 1966.

Goodrich, Harold B. "The Past and the Future." *American Association of Petroleum Geologists Bulletin* 5, no. 4 (1921): 45–57.

Goodwyn, Lawrence. *Texas Oil, American Dreams: A Study of the Texas Independent Producers and Royalty Owners Association*. Austin: Texas State Historical Association, 1996.

Gorman, Michael E. "Types of Knowledge and Their Roles in Technology Transfer." *Journal of Technology Transfer* 27 (2002): 219–31.

Gould, Charles Newton. "Beginning of the Geological Work in Oklahoma." *Chronicles of Oklahoma* 10 (1932): 196–203.

———. "Billion Barrels of Oil Still Lies beneath Surface of Oklahoma Soil." *Daily Oklahoman*, September 23, 1917.

———. *Covered Wagon Geologist*. Norman: University of Oklahoma Press, 1959.

———. "Geological Work in the Southwest." *Bulletin of the Southwestern Association of Petroleum Geologists* 1 (1917): 20–33.

———. *The Geology and Water Resources of the Western Portion of the Panhandle of Texas*. U.S. Geological Survey, Water Supply and Irrigation paper no. 191. Washington DC: GPO, 1907.

———. "Memorial: Lon Lewis Hutchison." *American Association of Petroleum Geologists Bulletin* 31 (March 1947): 650–51.

———. "Oklahoma: An Example of Arrested Development." *Economic Geography* 2, no. 3 (1926): 426–50.

———. *Oklahoma: The Geologist's Laboratory*. Oklahoma Geological Survey circular no. 16. Norman: Oklahoma Geological Survey, 1927.

———. "Petroleum and Natural Gas in Oklahoma." *Economic Geology* 7, no. 8 (1912): 719–31.

———. "Petroleum and Surface Vegetation." *Proceedings of the Oklahoma Academy of Science* 10 (1930): 110–14.

———. "Pioneer Geology in Oklahoma." *Tulsa Geological Society Digest* 14 (1945–46): 56–57.

———. *Travels through Oklahoma*. Oklahoma City: Harlow Publishing, 1928.

Gould, Charles Newton, and Frank E. Lewis. *The Permian of Western Oklahoma and the Panhandle of Texas*. Oklahoma Geological Survey circular no. 13. Norman: Oklahoma Geological Survey, 1926.

Gould, Charles Newton, L. L. Hutchison, and Nelson Gaylord. *Preliminary Report on the Mineral Resources of Oklahoma*. Oklahoma Geological Survey bulletin no. 1. Norman: Oklahoma Geological Survey, 1908.

Gressley, Gene. "GOS, Petroleum, Politics and the West." In *The Twentieth Century American West: A Potpourri*, edited by Gene Gressley, 102–38. Columbia: University of Missouri Press, 1977.

Hager, Dorsey. "The Evidence of the Oklahoma Oil Fields on the Anticlinal Theory." *Transactions of the American Institute of Engineers* 56 (February 1917): 843–55.

Hamilton, W. R. "Development of Petroleum Geology in California." *American Association of Petroleum Geologists Bulletin* 5, no. 4 (1921): 457–60.

Hanson, Maynard J. "Senator William B. Pine and His Times." Master's thesis, University of South Dakota, 1983.

Harper, John A. "The Incredible John F. Carll: The World's First Petroleum Geologists and Engineer." *Oilfield Journal*, Winter 2001.

Haskell, Thomas L. *The Authority of Experts: Studies in History and Theory*. Interdisciplinary Studies in History. Bloomington: Indiana University Press, 1984.

Hatzopoulos, John N. *Topographic Mapping: Covering the Wider Field of Geospatial Information Science and Technology (Gis&T)*. Boca Raton FL: Universal Publishers, 2008.

Hawkins, Hugh. "University Identity: The Teaching and Research Functions." In *Building the Organizational Society: Essays on Associational Activities in Modern America*, edited by Jerry Israel, 285–312. New York: Free Press, 1972.

Hayes, C. W. *Field Mapping for the Oil Geologist*. New York: Wiley, 1921.

———. *Handbook for Field Geologists*. New York: Wiley, 1913.

Hendrickson, Walter B. "Nineteenth-Century State Geological Surveys: Early Governmental Support of Science." *Isis* 52 (September 1961): 357–71.

Henry, J. T., comp. *Early and Later History of Petroleum*. Philadelphia: Jas. B. Rodgers, 1873.

Higham, John. "The Matrix of Specialization." In *Building the Organizational Society: Essays on Associational Activities in Modern America*, edited by Jerry Israel, 3–18. New York: Free Press, 1972.

Hineline, Mark L. "The Visual Culture of the Earth Sciences, 1863–1970." PhD diss., University of California–San Diego, 2002.

History of Petroleum Engineering. New York: American Petroleum Institute, 1961.

Hobson, Wayne K. "Professionals, Progressives, and Bureaucratization: A Reassessment." *The Historian* 39, no. 4 (1977): 639–58.

Hounshell, David A. "Hughesian History of Technology and Chandlerian Business History: Parallels, Departures, and Critics." *History and Technology* 12 (1995): 204–14.

Howell, Jesse V. "Historical Development of the Structural Theory." In *Problems of Petroleum Geology*, edited by W. E. Wrather and F. H. Lahee, 1–23. Tulsa: American Association of Petroleum Geologists, 1934.

Huffman, George Garrett, comp. *History of the School of Geology and Geophysics: The University of Oklahoma*. Norman: Alumni Advisory Council of the School of Geology and Geophysics, University of Oklahoma, 1990.

Hughes, Thomas P. *American Genesis: A Century of Invention and Technological Enthusiasm, 1870–1970*. New York: Penguin Books, 1989.

———. "The Electrification of America: The System Builders." In *The Engineer in America*, edited by Terry Reynolds, 191–228. Chicago: University of Chicago Press, 1991.

———. *Elmer Sperry, Inventor and Engineer*. Baltimore: Johns Hopkins University Press, 1971.

———. "Emerging Themes in the History of Technology." *Technology and Culture* 20 (October 1979): 697–711.

———. "The Evolution of Large Technological Systems." In *The Social Construction of Technological Systems*, edited by Wiebe E Bijker, Thomas P. Hughes, and Trevor J. Pinch, 51–82. Cambridge MA: MIT Press, 1987.

———. *Networks of Power: Electrification in Western Society, 1880–1930*. Baltimore: Johns Hopkins University Press, 1983.

———. "The Seamless Web: Technology, Science, Etcetera, Etcetera." *Social Studies of Science* 16 (1986): 281–92.

"Hunch, Backed by Unfailing Courage and Faith Resulted in Discovery of Pioneer Field." *Oil and Gas Journal* 21 (June 8, 1922): 32.

Ise, John. *United States Oil Policy*. New Haven CT: Yale University Press, 1926.

Jacoby, Karl. *Crimes against Nature: Squatters, Poachers, Thieves and the Hidden History of American Conservation*. Berkeley: University of California Press, 2001.

Johnson, Arthur Menzies. *Petroleum Pipelines and Public Policy, 1906–1959*. Cambridge MA: Harvard University Press, 1967.

Johnson, Carol Siri. "The Evolution of Illustrated Texts and Their Effect on Science: Examples from Early American State Geological Reports." *Leonardo* 41, no. 2 (2008): 120–27.

Johnson, Terence James. *Professions and Power*. Studies in Sociology. London: Macmillan, 1972.

Jones, W. Alton. *The Cities Service Story: A Case of American Enterprise*. New York: Newcomen Society in North America, 1955.

Karasick, Norman M., and Dorothy K. Karasick. *The Oilman's Daughter: A Biography of Aline Barnsdall*. Encino CA: Carleston Publishing, 1993.

Kimball, Bruce A. *The "True Professional Ideal" in America: A History*. Cambridge MA: Blackwell, 1992.

Knowles, Ruth Sheldon. *The Greatest Gamblers: The Epic of American Oil Exploration*. New York: McGraw-Hill, 1959.

Kohler, Robert E. *Landscapes and Labscapes: Exploring the Lab-Field Border in Biology*. Chicago: University of Chicago Press, 2002.

Kohlstedt, Sally Gregory. "The Geologists' Model for National Science, 1840–1847." *Proceedings of the American Philosophical Society* 118, no. 2 (1974): 179–95.

Kuhn, Thomas S. *The Structure of Scientific Revolutions*. Chicago: University of Chicago Press, 1996.

Kuklick, Henrika, and Robert E. Kohler, eds. *Science in the Field*. Special issue, *Osiris* 11 (1996).

Lacey, Michael J. "The World of the Bureaus: Government and the Positivist Project in the Late Nineteenth Century." In *The State and Social Investigation in Britain and the United States*, edited by Michael J. Lacey and Mary O. Furner, 127–70. Washington DC: Woodrow Wilson Center Press, 1993.

Lahee, Frederic H. "The Education of the Geologist." *Economic Geology* 19, no. 7 (1924): 684–86.

———. *Field Geology*. 2nd ed. New York: McGraw-Hill, 1923.

Landes, Kenneth K. *Petroleum Geology*. New York: Wiley, 1951.

Langston, Nancy. *Where Land and Water Meet: A Western Landscape Transformed*. Weyerhaeuser Environmental Book. Seattle: University of Washington Press, 2003.

Larson, Magali Sarfatti. *The Rise of Professionalism: A Sociological Analysis*. Berkeley: University of California Press, 1977.

Latour, Bruno. *Reassembling the Social: An Introduction to Actor-Network-Theory*. New York, Oxford University Press, 2005.

Laudan, Rachel *From Mineralogy to Geology*. Chicago: University of Chicago Press, 1993.

Lave, Jean, and Etienne Wenger. *Situated Learning: Legitimate Peripheral Participation*. Learning in Doing. New York: Cambridge University Press, 1991.

Law, John, and John Hassard. *Actor Network Theory and After*. Oxford: Blackwell/ Sociological Review, 1999.

Layton, Edwin T., Jr., "Mirror-Image Twins: The Communities of Science and Technology in 19th-Century America." *Technology and Culture* 12 (October 1971): 562–80.

———. *The Revolt of the Engineers: Social Responsibility and the American Engineering Profession*. Baltimore: Johns Hopkins University Press, 1986.

Lears, Jackson. *Something for Nothing: Luck in America*. New York: Penguin Putnam, 2003.

Lesley, J. Peter. "Early Observations of the Geology of Pennsylvania." In *Historical Sketch of Geological Explorations in Pennsylvania and Other States*, edited by J. P. Lesley, 3–28. Second Geological Survey of Pennsylvania. Harrisburg PA: Board of Commissioners for the Second Geological Survey, 1876.

———. "The Geology of the Pittsburgh Coal Region." *Transactions of the American Institute of Mining Engineers* 14 (June 1885–May 1886): 618–56.

———, ed. *Historical Sketch of Geological Explorations in Pennsylvania and Other States*. Second Geological Survey of Pennsylvania. Harrisburg PA: Board of Commissioners for the Second Geological Survey, 1876.

———. "A History of the First Geological Survey of Pennsylvania." In *Historical Sketch of Geological Explorations in Pennsylvania and Other States*, edited by J. P. Lesley, 53–134. Second Geological Survey of Pennsylvania. Harrisburg PA: Board of Commissioners for the Second Geological Survey, 1876.

———. *The Iron Manufacturers' Guide to the Furnaces, Forges, and Rolling Mills of the United States with Discussions of Iron as a Chemical Element, an American Ore, and a Manufactured Article, in Commerce and in History*. New York: Wiley, 1859.

———. *Manual of Coal and Its Topography*. Philadelphia: Lippincott, 1856.

———. "Petroleum in the Eastern Coal-Field of Kentucky." *Proceedings of the American Philosophical Society* 10 (April 1865): 33–69.

———. "Second Geological Survey of Pennsylvania [1874]." In *Historical Sketch of Geological Explorations in Pennsylvania and Other States*, edited by J. P. Lesley, i–xiv. Second Geological Survey of Pennsylvania. Harrisburg PA: Board of Commissioners for the Second Geological Survey, 1876.

———. "Second Geological Survey of Pennsylvania [1875]." In *Historical Sketch of Geological Explorations in Pennsylvania and Other States*, edited by J. P. Lesley, xv–xxvi. Second Geological Survey of Pennsylvania. Harrisburg PA: Board of Commissioners for the Second Geological Survey, 1876.

Leuchtenburg, William E. "The Pertinence of Political History: Reflections on the Significance of the State in America." *Journal of American History* 73 (December 1986): 585–600.

Levorsen, A. I. *Geology of Petroleum*. San Francisco: Freemon, 1959.

———. "Stratigraphic Versus Structural Accumulation." *American Association of Petroleum Geologists Bulletin* 20, no. 5 (1936): 521–30.

Levy, David W. "Scientist and Bard: The Poetry of Charles N. Gould." *Sooner Magazine*, Fall 1996.

———. *The University of Oklahoma: A History*. Norman: University of Oklahoma Press, 2005.

Limerick, Patricia Nelson. *Legacy of Conquest: The Unbroken Past of the American West*. New York: North, 1987.

Lively, Robert A. "The American System: A Review Article." *Business History Review* 29 (1955): 81–96.

Lucier, Paul. "Commercial Interests and Scientific Disinterestedness: Consulting Geologists in Antebellum America." *Isis* 86 (1996): 245–67.

———. "Scientists and Swindlers: Coal, Oil, and Scientific Consulting in the American Industrial Revolution." PhD diss., Princeton University, 1994.

———. *Scientists and Swindlers: Consulting on Coal and Oil in America, 1820–1890*. Baltimore: Johns Hopkins University Press, 2008.

Lyman, Benjamin Smith. "Biographical Notice of J. Peter Lesley." *Transactions of the American Institute of Engineers* 34 (1903): 726–39.

Mallison, Sam T. *The Great Wildcatter*. Charleston: Education Foundation of West Virginia, 1953.

Manning, Thomas G. *Government in Science: The U.S. Geological Survey, 1867–1894*. Lexington: University of Kentucky Press. 1967.

Mayr, Otto. "The Science-Technology Relationship as an Historiographic Problem." *Technology and Culture* 17 (1976): 663–73.

McDermott, Etienne, Richard Snyder, and William Wenger. *Cultivating Communities of Practice: A Guide to Managing Knowledge*. Boston: Harvard Business School Press, 2002.

McEvoy, Arthur F. "Working Environments: An Ecological Approach to Industrial Health and Safety." *Technology and Culture* 36, suppl. (1995): S145–S173.

McGerr, Michael. "Is There a Twentieth Century West?" In *Under an Open Sky: Rethinking America's Western Past*, edited by William Cronon, George Miles, and Jay Gitlin, 239–56. New York: Norton, 1992.

Meiksins, Peter. "The Revolt of the Engineers Reconsidered." *Technology and Culture* 29 (1988): 219–46.

Meinzer, O. E. Introductory Note. In *The Divining Rod: A History of Water Witching*, by Arthur J. Ellis, 5–6. U.S. Geological Survey, Water Supply paper 416. Washington DC: GPO, 1917.

Mellon, William Larimer. *Judge Mellon's Sons*. Privately printed by the author, 1948.

Merrill, George P. *The First One Hundred Years of American Geology*. New Haven CT: Yale University Press, 1924.

Merrill, Karen R. "In Search of 'the Federal Presence' in the American West." *Western Historical Quarterly* 30 (1999): 449–73.

Miles, Ray. *"King of the Wildcatters": The Life and Times of Tom Slick, 1883–1930*. College Station: Texas A&M University Press, 1996.

Millbrooke, Anne. "Henry Darwin Rogers and the First State Geological Survey of Pennsylvania." *Northeastern Geology* 3 (1981): 71–74.

Miller, Keith. "Petroleum Geology to 1920." In *Sciences of the Earth: An Encyclopedia of Events, People, and Phenomena*, edited by Gregory A. Good, 670–75. Vol. 2. New York: Garland, 1995.

Moran, Robert B. "The Role of the Geologist in the Development of the California Oil Fields." *American Association of Petroleum Geologists Bulletin* 8 (January–February 1924): 73–77.

Morris, Edmund. *Derrick and Drill, or An Insight into the Discovery, Development and Present Condition and Future Prospects of Petroleum in New York, Pennsylvania, Ohio and West Virginia*. New York: James Miller, 1865.

Morse, Kathryn Taylor. *The Nature of Gold: An Environmental History of the Klondike Gold Rush*. Seattle: University of Washington Press, 2003.

Nash, Gerald D. *United States Oil Policy, 1890–1964: Business and Government in the Twentieth Century*. Pittsburgh: University of Pittsburgh Press, 1968.

Nash, Linda. "The Changing Experience of Nature: Historical Encounters with a Northwest River." *Journal of American History* 86, no. 4 (2000): 1600–1629.

Newby, Jerry B. "Daniel Webster Ohern." *American Association of Petroleum Geologists Bulletin* 38, no. 5: 962–64.

————. "George Franklin Buttram and Fortuna Oil Company." In *History of the School of Geology and Geophysics, the University of Oklahoma*, edited by George Garrett Huffman, 245. Norman: Alumni Advisory Council of the School of Geology and Geophysics, University of Oklahoma, 1990.

Newell, Julie. "American Geologists and Their Geology: The Formation of the American Geological Community, 1780–1865." PhD diss., University of Wisconsin, 1993.

Noble, David. *America by Design: Science, Technology, and the Rise of Corporate Capitalism*. Oxford, UK: Oxford University Press, 1977.

Nye, David E. "Technology, Nature and American Origin Stories." *Environmental History* 8 (January 2003): 8–24.

Nystrom, Eric Charles. "Learning to See: Visual Tools in American Mining Engineering, 1860–1920." PhD diss., Johns Hopkins University, 2007.

Oldroyd, David R. "The Earth Sciences." In *From Natural Philosophy to the Sciences*, edited by David Cahan, 88–128. Chicago: University of Chicago Press, 2003.

Oleson, Alexandra, and John Voss. Introduction. In *The Organization of Knowledge in Modern America*, edited by Alexandra Oleson and John Voss, vii–xxi. Baltimore: Johns Hopkins University Press, 1979.

Olien, Roger M., and Diana David Olien. *Easy Money: Oil Promoters and Investors in the Jazz Age*. Chapel Hill: University of North Carolina Press, 1990.

————. *Oil and Ideology: The Cultural Creation of the American Petroleum Industry*. Chapel Hill: University of North Carolina Press, 2000.

————. *Oil in Texas: The Gusher Age, 1895–1945*. Austin: University of Texas Press, 2002.

————. *Wildcatters: Texas Independent Oilmen*. Austin: Texas Monthly Press, 1984.

Oreskes, Naomi. "Why Predict? Historical Perspectives on Prediction and in Earth Science." In *Prediction, Science, Decision Making and the Future of Nature*, edited by Daniel Sarewitz, Roger A. Pielke Jr., and Radford Byerly Jr., 23–40. Washington DC: Island Press, 2000.

Orr, Julian E. *Talking about Machines: An Ethnography of a Modern Job*. Ithaca NY: ILR Press, 1996.

Orton, James. *Underground Treasures: How and Where to Find Them*. Hartford CT: Worthington, Dustino, 1872.

Owen, Edgar. *Trek of the Oil Finders: A History of Exploration for Petroleum*. Tulsa: American Association of Petroleum Geologists, 1975.

Pandora, Katherine. "Knowledge Held in Common: Tales of Luther Burbank and Science in the American Vernacular." *Isis* 92 (September 2001): 484–516.

————. "Natural Histories of Science in the American Vernacular, or Knowing One's Place in the History of Science." Paper presented at the Andrew W.

Mellon Conference, "Where the Biological and Social Converge: Identity, Place, and Knowledge in the History of Science," University of Oklahoma, April 15–17, 1999.

Pauly, Philip J. "Science." In *A Companion to American Thought*, edited by Richard Wightman Fox and James T. Kloppenberg, 613–16. Oxford: Blackwell, 1995.

Peckham, Stephen Farnum. *Report on the Production, Technology, and Uses of Petroleum and Its Products*. Washington DC: GPO, 1885.

Pees, Samuel T. "Early Oil and Gas Exploration to 1879 in Western Crawford County, Pennsylvania." *Earth Sciences History* 2, no. 2 (1983): 103–21.

Pettengill, Samuel B. *Hot Oil: The Problem of Petroleum*. New York: Economic Forum, 1936.

Polanyi, Michael. *The Tacit Dimension*. Garden City NY: Doubleday, 1966.

Powers, Sidney. "Petroleum Geology in Oklahoma." In *Oil and Gas in Oklahoma*. Oklahoma Geological Survey bulletin 40. Vol. 1. Norman: Oklahoma Geological Survey, 1928.

Pratt, Joseph A. "The Petroleum Industry in Transition: Antitrust and the Decline of Monopoly Control in Oil." *Journal of Economic History* 40 (December 1980): 815–37.

Pratt, Wallace E. "Geology in the Petroleum Industry." *American Association of Petroleum Geologists Bulletin* 24 (July 1940): 1212–13.

———. *Oil in the Earth*. Lawrence: University Press of Kansas, 1942.

Price, Jennifer. *Flight Maps: Adventures with Nature in Modern America*. New York: Basic Books, 1999.

Price, Paul H. "Anticlinal Theory and Later Developments in West Virginia." *American Association of Petroleum Geologists Bulletin* 22 (August 1938): 1097–1100.

Prindle, David F. *Petroleum Politics and the Texas Railroad Commission*. Austin: University of Texas Press, 1981.

Pyne, Stephen J. *Fire in America: A Cultural History of Wildland and Rural Fire*. Princeton NJ: Princeton University Press, 1982.

———. *Grove Karl Gilbert: A Great Engine of Research*. Austin: University of Texas Press, 1980.

———. *How the Canyon Became Grand: A Short History*. New York: Viking, 1998.

———. *The Ice: A Journey to Antarctica*. Iowa City: University of Iowa Press, 1986.

Rabbitt, J. S., and Mary C. Rabbitt. "The U.S. Geological Survey: 75 Years of Service to the Nation, 1879–1954." *Science* 28 (May 1954): 741–58.

Rabbitt, Mary C. *The United States Geological Survey, 1879–1989*. U.S. Geological Survey circular. Washington DC: GPO, 1989.

Rae, John. "The Application of Science to Industry." In *The Organization of Knowledge in Modern America, 1860–1920*, edited by Alexandra Oleson and John Voss, 249–68. Baltimore: Johns Hopkins University Press, 1979.

Ransome, Frederick L. "The Present Standing of Applied Geology." *Economic Geology* 1 (October–November 1905): 1–10.

Reynolds, Terry S. "The Engineer in 19th-Century America." In *The Engineer in America*, edited by Terry S. Reynolds, 7–26. Chicago: University of Chicago Press, 1991.

Rister, Carl Coke. *Oil! Titan of the Southwest*. Norman: University of Oklahoma Press, 1949.

Rogers, Emma, and W. T. Sedgwick Rogers, eds. *Life and Letters of William B. Rogers*. 2 vols. Boston: Houghton Mifflin, 1896.

Rogers, Henry Darwin. *First Annual Report of the State Geologist*. Harrisburg PA: Emanuel Guyer, 1836.

Rogers, William Barton. *A Few Facts Regarding the Geological Survey of Pennsylvania: Exposing the Erroneous Statements and Claims of J. P. Lesley, Secretary of the American Iron Association*. Philadelphia: Collins, 1859.

Rose, Mark H. *Cities of Light and Heat: Domesticating Gas and Electricity in Urban America*. University Park: Pennsylvania State University, 1995.

Rudwick, Martin J. S. "The Emergence of a Visual Language for Geological Science, 1760–1840." *History of Science* 14, no. 3 (1976): 149–95.

———. "Geological Travel and Theoretical Innovation: The Role of 'Liminal.'" *Social Studies of Science Experience* 26 (1996): 143–59.

———. *The Great Devonian Controversy: The Shaping of Scientific Knowledge among Gentlemanly Specialists*. Chicago: University of Chicago Press, 1985.

Rürup, Reinhard. "Historians and Modern Technology: Reflections on the Development and Current Problems of the History of Technology." *Technology and Culture* 15 (April 1974): 161–93.

Sabin, Paul. *Crude Politics: The California Oil Market, 1900–1940*. Berkeley: University of California Press, 2005.

Santschi, Roy J. *Modern "Divining Rods": A History and Explanation of Geophysical Prospecting Methods, Including Descriptions of Instruments, and Useful Information for Prospectors and Treasure Seekers*. Glen Ellyn IL: Santschi, 1928.

Sarewitz, Daniel, Roger A. Pielke Jr., and Radford Byerly Jr. "Introduction: Death, Taxes, and Environmental Policy." In *Prediction: Science, Decision Making and the Future of Nature*, edited by Daniel Sarewitz, Roger A. Pielke Jr., and Radford Byerly Jr., 1–7. Washington DC: Island Press, 2000.

Scales, James R., and Danney Goble. *Oklahoma Politics: A History*. Norman: University of Oklahoma Press, 1980.

Schruben, Francis W. *From Wea Creek to El Dorado*. Columbia: University of Missouri Press, 1972.

"Scientific Methods of Exploration for Oil Help Industry Meet Its Demands." *Oil and Gas Journal* 34, no. 31 (December 19, 1935): 13–14.

Scott, James C. *Seeing Like a State: How Certain Schemes to Improve the Human Condition Have Failed.* New Haven CT: Yale University Press, 1998.

Sellars, Nigel Anthony. *Oil, Wheat, and Wobblies: The Industrial Workers of the World in Oklahoma, 1905–1930.* Norman: University of Oklahoma Press, 1998.

Sewell, William H., Jr. "The Concept(s) of Culture." In *Beyond the Cultural Turn: New Directions in the Study of Society and Culture,* edited by Victoria E. Bonnell and Lynn Hunt, 35–61. Berkeley: University of California Press, 1999.

Shapin, Steven. "Placing the View from Nowhere: Historical and Sociological Problems in the Location of Science." *Transactions of the Institute of British Geographers* 23, no. 1 (1998): 5–12.

———. *The Scientific Life: A Moral History of a Late Modern Vocation.* Chicago: University of Chicago Press, 2008.

Sherretts, William Bruce, and Joshua F. Moore. *Oil Boom Architecture: Titusville, Pithole, and Petroleum Center.* Charleston SC: Arcadia, 2008.

Shils, Edward. "The Order of Learning in the U.S." In *The Organization of Knowledge in Modern America, 1860–1920,* edited by Alexandra Oleson and John Voss, 19–50. Baltimore: Johns Hopkins University Press, 1979.

Silliman, Benjamin. "The Divining Rod." *American Journal of Science* 11, no. 2 (October 1826): 201–12.

Simon, John Y., ed. *The Papers of Ulysses S. Grant.* Vol. 27, *January 1 to October 31, 1876.* Carbondale: Southern Illinois University Press, 2005.

Singer, Jonathan W. *Broken Trusts: The Texas Attorney General Versus the Oil Industry, 1889–1909.* College Station: Texas A&M University Press, 2002.

Smith, Michael L. *Pacific Visions: California Scientists and the Environment, 1850–1915.* New Haven CT: Yale University Press, 1987.

Smith, Pamela H. "Science on the Move: Recent Trends in the History of Early Modern Science." *Renaissance Quarterly* 62 (2009): 345–75.

Sorensen, Knut H., and Nora Levold. "Tacit Networks, Heterogeneous Engineers, and Embodied Technology." *Science, Technology, and Human Values* 17 (Winter 1992): 13–35.

Spence, Clark C. *Mining Engineers and the American West: The Lace-Boot Brigade, 1849–1933.* New Haven CT: Yale University Press, 1970.

"State Geological Surveys and Economic Geology." *Economic Geology* 20 (June–July 1925): 376–81.

Staudenmaier, John M. *Technology's Storytellers: Reweaving the Human Fabric.* Cambridge MA: Society for the History of Technology and MIT Press, 1985.

Stevenson, John J. "J. Peter Lesley." *Science* 18, no. 444 (1903): 1–3.

———. "Memoir of J. Peter Lesley." *Bulletin of the Geological Society of America* 15 (1904): 532–41.

Stine, Jeffrey K., and Joel A. Tarr. "At the Intersection of Histories: Technology and the Environment." *Technology and Culture* 39, no. 4 (1998): 601–40.

Sutter, Paul. *Driven Wild: How the Fight against Automobiles Launched the Modern Wilderness Movement.* Seattle: University of Washington Press, 2002.

Sweet, George Elliott. *Gentleman in Oil.* Los Angeles: Science Press, 1966.

Tait, Samuel. *The Wildcatters: An Informal History of Oil Hunting in America.* Princeton NJ: Princeton University Press, 1946.

Turner, Frederick Jackson. *Significance of the Frontier in American History.* New York: Holt, 1920.

Tyrrell, Ian R. *True Gardens of the Gods: Californian-Australian Environmental Reform, 1860–1930.* Berkeley: University of California Press, 1999.

Tyson, Carl N., James Harold Thomas, and Odie B. Faulk. *The McMan: The Lives of Robert M. McFarlin and James A. Chapman.* Norman: University of Oklahoma Press, 1977.

Valencius, Conevery Bolton. *The Health of the Country: How American Settlers Understood Themselves and Their Land.* New York: Basic Books, 2002.

Veysey, Laurence R. *The Emergence of the American University.* Chicago: University of Chicago Press, 1965.

———. "Who's a Professional? Who Cares?" *Reviews in American History* 3, no. 4 (1975): 419–23.

Warner, Charles A. "Sources of Men." In *History of Petroleum Engineering,* 35–61. New York: American Petroleum Institute, 1961.

Warren, Louis S. *The Hunter's Game: Poachers and Conservationists in Twentieth-Century America.* New Haven CT: Yale University Press, 1997.

Weber, Max. *Economy and Society: An Outline of Interpretive Sociology.* Edited by Guenther Roth and Claus Wittich. 2 vols. Berkeley: University of California Press, 1978.

Wengenroth, Ulrich. "Science, Technology, and Industry." In *From Natural Philosophy to the Sciences: Writing the History of Nineteenth Century Science,* edited by David Cahan, 221–53. Chicago: University of Chicago Press, 2003.

White, Gerald Taylor. "California's Other Mineral." *Pacific Historical Review* 39, no. 2 (1970): 135–54.

———. *Formative Years in the Far West; a History of Standard Oil Company of California and Predecessors through 1919.* New York: Arno, 1976.

———. "Oil Industry." In *The Reader's Encyclopedia of the American West,* edited by Howard Lamar, 860–65. New York: Harper & Row, 1977.

White, Israel C. "The Geology of Natural Gas." *Science* 6 (June 26, 1885): 521–22.

———. "The Mannington Oil Field and the History of its Development." *Geological Society of America Bulletin* 3 (1892): 187–216.

White, Richard. "'Are You an Environmentalist or Do You Work for a Living?': Work and Nature." In *Uncommon Ground: Rethinking the Human Place in Nature*, edited by William Cronon, 171–85. New York: Norton, 1995.

———. "From Wilderness to Hybrid Landscapes: The Cultural Turn in Environmental History." *The Historian* 66, no. 3 (2004): 557–64.

———. *"It's Your Misfortune and None of My Own": A History of the American West.* Norman: University of Oklahoma Press, 1991.

———. *The Organic Machine: The Remaking of the Columbia River.* New York: Hill and Wang, 1995.

Wiebe, Robert H. *The Search for Order, 1877–1920.* New York: Hill and Wang, 1967.

Worster, Don. *Rivers of Empire: Water, Aridity, and the Growth of the American West.* Oxford, UK: Oxford University Press, 1992.

Wright, William. *Oil Regions of Pennsylvania, Showing Where Petroleum Is Found, How It Is Obtained, and at What Cost, with Hints for Whom It May Concern.* New York: Harper, 1865.

Wrigley, Henry E. "Present and Future of the Pennsylvania Oil Fields." *Engineering and Mining Journal* 28 (July–December 1879): 186.

Wrigley, Henry E., Lucas D. Jones, and J. Peter Lesley. *Special Report on the Petroleum of Pennsylvania: Its Production, Transportation, Manufacture and Statistics.* Geological Survey of Pennsylvania, report of progress. Harrisburg PA: Board of Commissioners for the Second Geological Survey, 1875.

Yergin, Daniel. *The Prize: The Epic Quest for Oil, Money, and Power.* New York: Simon & Schuster, 1991.

Young, Otis E. *Western Mining: An Informal Account of Precious-Metals Prospecting, Placering, Lode Mining and Milling on the American Frontier from Spanish Times to 1893.* Norman: University of Oklahoma Press, 1970.

Index

Page numbers in italics refer to illustrations.

actor-network theory, 181n31
Amarillo Oil Company, 121
Amarillo TX, 120, 121
American Association of Petroleum
  Geologists, 106
Angell, Cyrus D., 40–42, 70, 186n106
anthracite coal, 57
anticlinal theory, 71–73, 81, 84,
  89, 119–22, 135, 137–39, 167–69,
  190nn120–21; acceptance of, 101,
  195n108; and Charles N. Gould,
  110–11
anticlines, 38–39, *71*, *72*, 82, 135, *161*;
  and beauty, 113; exhibit model of,
  138; mapping of, 166; in Texas Panhandle, 120. *See also* Augusta anticline; Coalinga anticline; Cushing
  anticline; El Dorado anticline
antitrust ruling. *See* Standard Oil
  Company
Appalachian Mountains, 56
Arbuckle Mountains, 112–13, *113*, *114*,
  115, *116*, 124

Arnold, Ralph, 90, 197n80
asphalt veins, 24
Associated Oil Company, 125
Augusta anticline, 160–61
Augusta oil field KS, 117, 159–61, 163,
  171, 195n108
authority, 2, 117–18, 172; and early
  prospecting methods, 44; and
  geological theories, 46, 68–69,
  121–22, 174; and geologists, 43,
  103–6, 122; institutional, 116, 139;
  intellectual, 61; internalized, 105;
  and local knowledge, 82; and map-
  making ability, 125; professional,
  58–61, 68–69, 104; and publicity,
  117–18; and reading the landscape,
  88; shift in, 46; and topographical
  science, 56

Barnsdall, Theodore N., 153–54, *154*,
  156, 201n43, 202n59
Barnsdall, William, 153–54
Barnsdall Oil Company, 154
Bartlesville OK, 81, 89, 152, 192n27
Beaumont TX, 103
Beebe, B. W., 119

Lears, Jackson, 25
Lesley, J. Peter, 52–56, 52, 58–70, 77, 169, 176
Levorsen, Arville I., 168–69
Lewis, James O., 102
local knowledge, 8, 42, 44–45, 94–98, 182n10, 182n13, 186n2; and authority, 82; and colonization, 23; and geological principles, 69; and geological theories, 75, 84; and geologists, 52–53, 56, 62–64, 77; and Pennsylvania, 74–77; and science, 15–16, 180n24; and southern plains, 82–83
luck, 25–26, 94, 96–97, 183n27. *See also* chance
Lyman, Benjamin, 64–66

Madame Virginia. *See* Bryan, Ruth
Magnolia Oil Company, 137
mana, 35
maps, 52, 58–60; of anticlines, 166; and authority, 125; contour, 73–74, 76, 76, 100, 120, 161, 164–66; and Empire Gas and Fuel Company, 159–61; and nature, 77; of Oklahoma energy development, 109; subsurface, 73–74, 76, 76, 100, 114, 194n94; topographical, 55–58, 77, 124, 127, 159–60, 162, 164–66, 188n46; and United States Geological Survey, 123, 127, 164–65
Marland, Ernest W., 92
McCoy, Alex W., 170–72
McDowell, J. C., 157
McMan Oil Company, 101
Melville, Herman: *The Confidence Man*, 26
Mercer County PA, 67

metis, 93, 95, 98, 193n60
Metropolitan Life Insurance Company, 136
Monongahela PA, 63

natural resources, 2, 4–5
nature: and authority, 117; and belt-line theory, 70; contested understandings of, 82–83, 175; and culture, 4, 116; different ways of knowing, 95; and geologists, 46, 77; interpretations of, 117; and local knowledge, 44, 77; and metis, 98; and practical oil men, 2, 4, 6–7, 44; and prospecting, 69; and students, 114; and topographical maps, 58, 77; and work, 6–7
Newell, Frederick, 127
New Jersey, 51
Newkirk OK, 110
New Mexico, 127
New York State Survey, 50
Noble County OK, 86
Nobles, Millard C., 121

Ohern, Daniel W., 131, 132, 133, 136–37
Oil and Gas Journal, 81
oil companies: and consultants, 161; foreign, 202n59; and geologists, 84, 111, 143, 161, 163–67, 164, 203nn94–95; and geology departments, 21; integrated, 172, 174
oil industry: and Doherty School of Practice, 151–52; and geologists, 82, 105; as monolith, 174–75; origins of, 24; and petroleum geology, 153, 161; and power, 174–75; and production practices, 83; and science, 104, 125–26; and United States Geological Survey, 124

Oil on the Brain Songster, 26
oil seers, 29–30. *See also* charismatics
oil smellers, 29, 85. *See also* charismatics
Oklahoma, 1, 13, 36, 108–9, *110*, 121. *See also* southern plains
Oklahoma Geological Survey, 106, *111*, 126–34, *132*, 198n106
Oklahoma Geological Survey Commission, 130
Oklahoma State Fair, *111*, *138*
overproduction, 14, 83, 177
Owen, Edgar, 161
Owen, Robert L., 128

Pennsylvania, 56–61, 72–77, 83; confidence men in, 26–27; geological surveys of, 12; and Herbert R. Straight, 156–57; landscape survey in, 47–48; oil in, 89, 176; and oil industry origins, 24; and William Barnsdall, 153–54
Pennsylvania Geological Survey, 50–55, 61, 65–66, 74–75
Penrose, Evelyn, 29
petroleum geology. *See* geology
Pew, J. Edgar, 173
Pine, William B., 154, 156
Pioneer oil field TX, 92
Polanyi, Michael, 33
politics, 50, 51, 55, 61–62, 129–30, 187n15
Powell, John Wesley, 123, 125
power, 2–3, 172; economic, 118; institutional, 106, 128, 139; and oil industry, 174–75; and professionalism, 15, 104, 112
practical oil men, 9–10, 12–13, 23, 35–42, 82–88, 174, 177, 192n43;

and craft tradition, 148–49; and Doherty School of Practice, 153; and gas utility business, 148; and geological theories, 97; and geologists, 75, 90–91, 101–2, 103, 133; and geology, 125; and Henry Chance, 69; methods of, 185n86; and nature, 2, 4, 6–7, 44; and reading the landscape, 102; and tacit knowledge, 91; and topographical maps, 159–60; and university-educated engineers, 149, 200n19
Pratt, Wallace, 91–92, 94, 96, 167, 183n27
Prentice, Frederick, 186n106
professionalism, 14–15, 105
prospecting: and geological principles, 39, 185n97; and geological structures, 100–102; and geological theories, 71, 119; and geology, 125; and intuitive knowledge, 96; and luck, 94, 96–97; and nature, 69; as sensory experience, 24, 44, 75, 85, 182n15; and speculation, 67–68; and surface topography, 169; and universities, 104
prospectors, 2, 9; and authority, 2; and geological processes, 4; and landscape, 175–76; methods of, 24–25, 74, 183n23; as pioneers, 91–92; vernacular, 23, 27–28. *See also* practical oil men

Quapaw Gas Company, 153

rationalism, 94–95
record keeping, 75. *See also* drillers' logs
Reeds, Arthur, *132*, *133*

reports, geological, 48–51, 58–61,
65–66
reservoirs, 39–40
Robinson, F. M., 156
Rockefeller, John D., 2, 14, 83, 174
Rogers, Henry D., 51, 55, 58–63
Rogers, William, 51, 55, 58–60
Roxana Petroleum Company, 137

Salt Creek WY, 123
sawmills, avoidance of, 39
science: abstract, 66; government-
sponsored, 125–26; and local
knowledge, 15–16, 180n24; and oil
industry, 22, 104, 118, 181n4, 181n6;
topographical, 55–58, 188n46
scientists, 45, 68, 75–76, 98, 120,
182n11, 186n1
seeps, 24
seismic waves, 177
Shaffer, Charles B., 98–100
Sidwell, J. S., 89
Silliman, Benjamin, 31
Sinclair Oil Corporation, 115
Slick, Tom, 13, 85–88, *85, 87*, 92, *99*,
174, 192n43; and Cushing oil field,
98–101; and gushers, 176; and Lon
Lewis Hutchison, 134
Slick, Tom, Jr., 88
Smith, Uncle Billy, 40
Snider, Luther C., 165
southern plains, 81–83, 89–90, 103,
143. *See also* Oklahoma; Texas
South Penn Oil Company, 89
Sperry, Elmer, 180n19
Spindletop oil field TX, 103, 157
Standard Oil Company, 14, 83, 84,
125, 143, 173–74, 200n2
Stanford University, 108, 156

Straight, Herbert R., 155, 156–57,
202n57, 202n59
strata, 60, 73, 82
stratigraphic traps, 73, *73*, 168–69
stratigraphy, 76–77, 114
students, 111–14, *115, 116, 126*, 131, 134,
198n106
subsurface exploration, 169–71,
205n124
subsurface maps, 73–74, 76, *76*, 100,
114, 194n94
supernaturalism, 28, 44, 191n19
superstition, 38–39
surveying equipment, 123, *124, 128*,
160, 165
surveys, geological, 49–55, 61, 126–30
synclines, 39
system-builders, 144–45, 174, 200n4

Taff, Joseph A., 127
Tarr, William A., 115
Taylor, Charles H., 117
Texas, 1, 120, 127. *See also* southern
plains
Tiger oil well OK, 99
Tishomingo OK, 121
Tonkawa oil boom, 86
topographical maps, 55–58, 77, *124*,
127, 159–60, *162*, 164–66, 188n46
topographical science, 55–58, 188n46
training schools, 147, 150, 152. *See also*
Doherty School of Practice
Tulsa OK, 173
Turner, Frederick Jackson, 36

Union des Pétroles d'Oklahoma,
202n59
Union Oil Company, 21, 125
United States Bureau of Mines, 126